Studies in the Methodology of Science

Igor Hanzel

Studies in the Methodology of Science

Bibliographic Information published by the Deutsche Nationalbibliothek
The Deutsche Nationalbibliothek lists this publication in the Deutsche Nationalbibliografie; detailed bibliographic data is available in the internet at http://dnb.d-nb.de.

Library of Congress Cataloging-in-Publication Data
Names: Hanzel, Igor.
Title: Studies in the methodology of science / Igor Hanzel.
Description: Frankfurt am Main : Peter Lang, 2016. | Includes bibliographical references.
Identifiers: LCCN 2015030939 | ISBN 9783631665114
Subjects: LCSH: Science—Philosophy. | Science—Methodology.
Classification: LCC Q175 .H286 2016 | DDC 501—dc23 LC record available at http://lccn.loc.gov/2015030939

ISBN 978-3-631-66511-4 (Print)
E-ISBN 978-3-653-05783-6 (E-Book)
DOI 10.3726/978-3-653-05783-6

© Peter Lang GmbH
Internationaler Verlag der Wissenschaften
Frankfurt am Main 2016
All rights reserved.
Peter Lang Edition is an Imprint of Peter Lang GmbH.

Peter Lang – Frankfurt am Main · Bern · Bruxelles · New York · Oxford · Warszawa · Wien

All parts of this publication are protected by copyright. Any utilisation outside the strict limits of the copyright law, without the permission of the publisher, is forbidden and liable to prosecution. This applies in particular to reproductions, translations, microfilming, and storage and processing in electronic retrieval systems.

This publication has been peer reviewed.

www.peterlang.com

Contents

Acknowledgments 9

Introduction 11

Chapter 1: Methodological Issues in Scientific Explanation 13
1.1 J. Woodward on Scientific Explanation 14
1.2 Conditions in Laws and Explanations 17
1.3 Heuristic in Scientific Explanation 22
1.4 Open Problems 26

Part I: Methodological Issues in Physics

Chapter 2: Bohr's Atom: Data, Phenomena, Laws of
Phenomena, and Explanations *via* Mechanism 31
2.1 Meta-reflections: Pure Metaphysics, Pure Epistemology,
and Epistemology-*cum*-Metaphysics 32
2.2 J. Bogen and J. Woodward on Data and Phenomena 35
2.3 From the Spectra of Gases to the Internal Motion in the
Molecules of Gases and "Back": The "Stoney" Program 40
2.4 The Spectra and the Stability of the Atom 44
    2.4.1 From data to phenomena: A.-J. Ångström 44
    2.4.2 From Phenomena to Laws of Phenomena:
    J. J. Balmer, J. R. Rydberg, and W. Ritz 50
    2.4.3 Towards the Atom: Its Stability and Its Electrons 56
2.5 Bohr's Stationary States and the Stability of the Atom 61
2.6 The Spectra Explained and Predicted 64
2.7 General Epistemological Lessons 71

Chapter 3: Observability and Theory-Ladenness of
Observation: Myths and Facts 79
3.1 Logical Positivism/Empiricism and the
Post-positivistic Backlash: The Myths 79
    3.1.1 From Logical Positivism to Logical Empiricism 79
    3.1.2 Post-positivism and the Theory Ladenness of Observation 84
3.2 Can a Theory-Loaded Theory Be Tested? A Case Study 85

    3.2.1 Balmer's Formula and Bohr's Hydrogen Atom 86
    3.2.2 An Attempt at an Epistemological Generalization 90
3.3 An Epistemological Way Out 94

Chapter 4: Kantian and Post-Kantian Themes in Early Quantum Mechanics 101

4.1 Kant on Intuition, Appearance, the Thing-in-Itself, and Categories 102
4.2 The Early Matrix Mechanics and Wave Mechanics 105
4.3 The Myth of Observability: Kantian Themes in Early Quantum Mechanics 110
4.4 Clarifications: Some Post-Kantian Themes in Early Quantum Mechanics 118

Chapter 5: Measurement and Conceptual Networks in Early Thermodynamics 125

5.1 Fire, Heat, Temperature, Thermometer and Weight × Distance 125
    5.1.1 Fire/heat and temperature 125
    5.1.2 Bootstrapping in the measurement of heat by means of the measurement of temperature 131
    5.1.3 Work as weight × distance 132
    5.1.4 Joule's on the Mechanical Equivalent of Heat 133
5.2 A Philosophical Explication 134
5.3 A Test: Thomson's Concept of Absolute Temperature 138

Part II: Methodological Issues in Primate Research

Chapter 6: The Methodological Turn in Ape Research: Sue Savage Rumbaugh 143

6.1 The Starting Points: The Late Sixties, Early Seventies, and the Lana Project 144
6.2 The Sherman-Austin Project 148
6.3 The Kanzi Project 156
6.4 Metascience and Methodology: From Behaviorism to Narrative Ethnography 162
6.5 Some Objections 169

Chapter 7: Varieties of Intentionality: Michael Tomasello 175

7.1 The Scientific and Meta-scientific Dimensions 175
    7.1.1 First order and second order intentionality:
        the eighties and nineties 175
    7.1.2 From first order and second order intentionality
        toward shared intentionality: the new millennium 180
    7.1.3 Shared Intentionality: from individual intentionality via
        joint intentionality to collective intentionality 185
7.2 The Methodological Dimension:
    Explanantia and Explananda 186
    7.2.1 Davidson, Krüger, Habermas, and the explanantia/explananda
        in the developmental and comparative psychology 186
    7.2.2 The structural, the structural-genetical, the structural-
        historical, and the historical-genetic methods in
        Tomasello's explanation of cooperative communication 188

Chapter 8: Tool use by Chimpanzees (*Pan Troglodytes*): New Conceptualization and a New Measure for Quantification 203

8.1 Introduction 203
8.2 Tool Use in Chimpanzees 203
    8.2.1 Leaf sponging 203
    8.2.2 Honey extracting 204
    8.2.3 Termite fishing 204
8.3 Chimpanzee Food Sharing 204
8.4 The Matsuzawa Model 205
8.5 The Hayashi – Mizuno – Matsuzawa Model 206
8.6 A New Conceptualization 207
8.7 A New Measure for Chimpanzee Tool Use 211

Part III: Methods of Theory Construction in Political Economy

Chapter 9: Marx's Method of Theory Construction: Categories, Magnitudes and Laws 215

9.1 Joan Robinson on Marx's Concepts of Value and on
    the Relation of Volume I to Volume III of *Capital* 216
9.2 Leszek Nowak on Marx's Method 218

9.3 The Methodological Implications of Hegel's
     Science of Logic for Marx 221
9.4 Marx's Methods in Volume I and in the Manuscripts of
     Books II and III of *Capital* 225
     9.4.1 *Capital* Volume I 225
     9.4.2 Manuscripts of Book II 236
     9.4.3 Manuscripts of Book III 239
9.5 Methodological Conclusions 242

Chapter 10: Adam Smith's Method of Theory Construction in Book I of *Wealth of Nations* 247

10.1 Exchangeable Value and Its Measure in Chapter V 248
10.2 Smith's Measure of Value and Econometrics 252
10.3 Measure, Standard and Cause: Meta-conceptual Reflections 255
10.4 Conclusion: Concepts, Categories and Smith's Natural Price 258

Chapter 11: Open Problems: Categories, Types of Thought Objects and the Historical Method 263

11.1 A Typology of Thought Objects for Types of Scientific Explanation 263
11.2 The Limits of the Applicability of the Category Cluster
      Appearance – Essence (Ground) – Manifestation 267

References 273

Acknowledgments

Several persons were helpful in the writing of this book. My colleagues from the Department of Logic and Methodology of Science at Comenius University read the drafts of several chapters of this book and suggested many improvements. I am especially grateful to Professor Roman Ciapalo from Loras College, Dubuque, Iowa, who over the years read the drafts of the book and made numerous suggestions to improve them. Without his support this book would never appear in print.

The Slovak version of Chapter 1 was published in the supplement of the journal *Filosofický Časopis*, 2015, Vol. 63. Chapters 2 and 6 were published in the journal *Organon F*, 2012, Vol. 19, No. 2, pp. 201–226 and 2013, Vol. 20, No. 3, pp. 302–322. Chapter 9 was published in the journal *International Critical Thought*, 2015, Vol. 5, No. 4. I am also grateful to Dr. J. Halas for letting me publish as Chapter 10 a paper on Adam Smith we authored together.

I am grateful to John Wiley & Sons publisher for the permission to reproduce two figures from the article (Savage-Rumbaugh 1981), pages 46 and 47, as well to MIT Press for the permission to publish four figures from the book (Tomasello 2008), pages 98, 105, 235, and 239.

Work on Chapter 1 was supported by the *Slovak Research and Development Agency* under contract No. APVV-0149-12. Work on Chapters 3, 4, 5, 7, 8, 9 10 and 11 was supported by the VEGA grant, grant number 1/0221/14.

Introduction

This book is a study of methods employed in the natural and social sciences. My approach to natural and social sciences is based, drawing partially on (Habermas 1981), on the view that one can discern in them a *practico-conceptual* dimension, that is, the dimension where they practically encounter nature and/or society and where they try to treat certain problems which they face in nature and society by intervening into nature and/or society based on the conceptualization of these problems. In addition to this dimension, I identify two more features: a *metaconceptual* dimension, where the choice of certain concepts in the first dimension is subjected to a specific reflective endeavor and a *methodological* dimension, where methods of derivation of concepts in the conceptual dimension for the purposes of, for example, explanation and prediction, employed are subjected to a special analysis.

This study is therefore accomplished by drawing on the following three resources. The first is the contemporary philosophy of science which deals with the methods of construction of conceptual systems in natural and social sciences. The second source is the epistemology of Kant's *Critique of Pure Reason* as well as the philosophical categories in Hegel's *Science of Logic*. I realize that it is highly unusual, to say the least, to draw on Hegel in the field of philosophy of science but as try to show in this book, the employment of these categories enables one to enlarge the conceptual framework of philosophy of science. The third source, functioning here as a "counterbalance" and as a testing ground for the first two, are various natural and social sciences.

As to the natural sciences, I choose pre-quantum spectral analysis, Bohr's quantum theory of the atom of 1913, the early quantum mechanics of Heisenberg and Schrödinger, and the beginnings of classical thermodynamics. As to the social sciences, I choose modern research into linguistic capabilities, cognition and instrumental action of great apes and human infants, as well as the economic theories of Marx and Adam Smith.

Chapter 1 focuses on the issue of scientific explanation vis-à-vis the work of J. Woodward, and provides a differentiated typology of scientific explanation as well as of heuristics involved therein. In Chapter 2 I focus on the reconstruction of epistemic categories and methods employed in modern spectral analysis as well as in Bohr's theory of the atom. Chapter 3 deals with the epistemological issue of observability and theory-ladenness of observation from the point of view of theories treated in Chapter 2. Chapter 4 approaches early quantum mechanics both in its matrix and wave forms from the point of view Kant's epistemology and shows that the very nature of this physical theory requires to pass over to a post-Kantian understanding which draws on realistico-epistemological interpretation of categories in Hegel's *Science of Logic*. Chapter 5 reconstructs the development of early classical thermodynamics, especially, the epistemological basis of the differentiation between the concept of heat and the concept of temperature, as well as the place of measurement procedures in this differentiation.

Chapter 6 deals with the methods of research into the cognitive capabilities of non-human great apes as conducted by S. E. Savage-Rumbaugh. Chapter 7's focus is on M. Tomasello's research into the structure of intentionality of both non-human great apes and human infants. Chapter 8 provides a conceptual reconstruction of the tool use by chimpanzees together with a new measure for the quantification of this use. Chapters 9 and 10 deal with the epistemico-categorial dimension of Marx's economic works and of Adam Smith's Book I of *Wealth of Nations*.

Finally, Chapter 11 delineates the questions and problems associated with the methodological investigation into the natural and the social sciences whose resolution lies in the future.

Chapter 1: Methodological Issues in Scientific Explanation

The aim of this chapter is to offer a concept of scientific explanation which draws on the works of J. Woodward starting from the late 1970s.[1] These works present in many respects an approach to the issue of scientific explanation which should be integrated into any reasonably serious philosophical-cum-methodological reconstruction of scientific explanation.

However, one has to bear in mind that Woodward presented certain aspects of his approach in a rather unspecified manner and, in addition, did not develop certain issues pertaining to scientific explanation with sufficient depth. My aim is to provide a remedy to these deficits.

In order to prevent any possible misunderstanding, I would like to emphasize that this chapter deals neither with Woodward's view on the counterfactual aspect of explanation, nor with his approach to issues of causation and invariance and their respective places in his reconstruction of scientific laws and explanations.[2]

I shall start with an overview of Woodward's approach to scientific explanation, namely, his differentiation between the (f) and (f') requirements for a valid explanation, his differentiation between the explanation of a law and of a singular phenomenon, and his requirement of a reconstrual of the *explanandum* in the course of scientific explanation. Then, I shall distinguish between modification conditions stated in scientific laws and singular conditions that are added in the course of scientific explanation to scientific laws from the outside, so to say. This differentiation will, then, enable me to distinguish methodologically between the explanation of a law and that of a singular phenomenon. Finally, I delineate some open problems to be solved in the future.

---

1   Woodward (1979; 1980; 1984; 1997; 2000b; 2003) and (Woodward and Hitchcock 2003).
2   On this see also (Hitchcock and Woodward 2003); for its analysis see (Strevens 2008), (Ylikoski and Kuorikoski 2010), (Imbert 2013) and (Saatsi and Pexton 2013).

1.1 J. Woodward on Scientific Explanation

In order to provide a view of scientific explanation which differs from that given in the D-N model, Woodward uses the following examples. First, he offers the following argument (1979, 41; 2003, 187):

> All ravens are black
> a is a raven (1.1)
> a is black

Then, he states cases of explanation encountered in classical mechanics and electrostatics. In the former (1979, 42), based on Newton's laws of motion and gravity, and the assumption that the Earth with mass M and radius R is a sphere and that the only force acting on a body with mass m falling from height h above the surface of the Earth is due to Earth's gravity, one obtains for the force acting on this body $F = G\dfrac{mM}{(R+h)^2} = ma$, where G is the gravitational constant and a is the acceleration acquired by the body. And, under the additional supposition that the height from which the body falls is much smaller than the radius of the Earth ($h \ll R$), one obtains for the acceleration of the falling body:

$$a = G\frac{M}{R^2} \quad (1.2)$$

From (1.2), in turn, it is possible by substituting the actual numerical values for G, M and R to explain the actual acceleration of a body falling freely above the Earth's surface.

In electrostatics, based on Coulomb's law, it is possible to derive the law stating the relation between the intensity of the electric field E at a perpendicular distance r from a very long fine wire on which is uniformly distributed a positive charge; this relation is (where λ stands for the charge per unit of length of the wire; ε_0 is a constant) (1979, 42; 1997, 27; 2003, 187):

$$E = \frac{1}{2\pi\varepsilon_0}\frac{\lambda}{r} \quad (1.3)$$

From Coulomb's law it is also possible to derive a law which states the intensity of electric field outside a uniformly charged hollow sphere of diameter r (where Q stands for the total charge of the sphere) (1997, 27):

$$E = \frac{1}{4\pi\varepsilon_0}\frac{Q}{r^2} \quad (1.4)$$

Woodward views the derivation of (1.2), of the actual value of a, as well as of (1.3) and (1.4), in contradistinction to (1.1), as instances of scientific explanation and characterizes the generalization figuring in these explanations as follows (1979, 47):

> These generalisations contain variables or parameters (mass, distance, acceleration ... charge, electrical intensity and so forth) which are such that a whole range of different states or conditions can be characterised in terms of variations of their values. The laws [involved in the explanations of (1.2), (1.3) and (1.4) ...] formulate a systematic relation between these variables. They show us how a range of different changes in certain of these variables will be linked to changes in others of these variables. In consequence, these generalisations are such that when the variables in them assume on set of values (when we make certain assumptions about boundary and initial conditions) the explananda in the above explanations are derivable, and when the variables in them assume other sets of values, a range of other explananda are derivable. For example, the second law of motion and the law of gravitation which occur in explanation [of (1.2) ...] are such that when the variables in them assume different values (*via* combinations of these generalisations with a different set of initial or boundary conditions) quite different explananda are derivable. For example, these generalisations are such that on the assumption that the mass and radius of the earth had different values, a quite different value for acceleration of a falling body could be derived. These generalisations are also such that we could use them to derive an expression for the rate of fall of body from a distance which is no longer negligible in comparison with the earth's radius. Indeed these generalisations are such that we could us them to derive even more disparate explananda; for example we could use them in conjunction with other information to derive Kepler's laws and a great many other derivative laws of Newtonian mechanics ... And in a similar way the version of Coulomb's law occurring in [explanation of (1.4) ...] can be used, as the parameters in this law assume different values, to explain a range of different explananda—the expression for electric intensity along the axis of a uniformly charged ring, or between two equally an oppositely uniformly charged plates, or inside and outside a uniformly charged hollow sphere, for example.

Based on this characterization, he imposes on explanations of both laws and singular *explananda* the *requirement of functional interdependence*, which goes as follows (1979, 46):

> (*f*) The law occurring in the explanans of a scientific explanation of some explanandum E must be stated in terms of variables or parameters variations in the values of which will permit the derivation of other explananda which are appropriately different from E.

At the same time he tries to specify the meaning of the phrase "variations in the value of the variable" by stating a modified requirement of functional interdependence which he does *not* accept (1979, 47):

15

(f′) The law occurring in the explanans of a scientific explanation of some explanandum E must be such that in conjunction with some appropriate set of initial and boundary conditions, it can be used to derive an explanandum which is appropriately different from E.

With respect to the aims of this chapter, two additional aspects of Woodward's approach to scientific explanation are worth mentioning. First, according to him, "a successful scientific explanation ... exhibits the explanandum in a new light, allowing one to see the relevance of certain considerations which were not apparent from the original characterization of the explanandum ... I shall express this by saying that a scientific explanation typically involves a 'reconstrual' of the explanandum" (1979, 61–62).

Second, even if he imposes the (f) requirement on the explanation of both laws and singular facts, still he emphasizes that "scientific explanations typically have as their explananda generalisations rather than singular sentences ... the scientific explanation of particular facts is an activity which is derivative or parasitic on the scientific explanation of generalisations" (1979, 63).

From this overview of Woodward's approach to scientific explanations I draw the following conclusions.

First, while Woodward tries – as can be seen from his differentiation between the (f) and (f′) requirements – to differentiate between, on the one hand, initial and boundary conditions figuring in the explanans of an explanation from, on the other hand, conditions relevant for explanation which differ somehow from the former, he does not characterize the latter. Instead, he speaks only very generally about "conditions cited in ... explanans" (1997, 25), and, when dealing with examples of explanation in classical mechanics and electrostatics, he in fact relapses into talk about "certain assumptions about boundary and initial conditions" (1979, 47).

Second, even when he emphasizes the priority of explanation of laws, as compared to that of singular facts, still he provides no foundations on the basis of which he could methodologically clarify the differences between these two types of scientific explanation. This was stressed, for example, by M. Strevens in his review of Woodward's *Making Things Happen* as follows: "the reader is for the most part left to extract the rules of generalization explanation from examples" (Strevens 2007, 236).

Finally, while Woodward views innovation as being an integral part of explanation, in the course of which the explanandum is transformed from the form it had initially – as an impulse and incentive for the attempts to explain it – to the form it acquires once derived by the explanation from the explanans, nevertheless he does not provide a detailed clarification what this innovation consists of.

The requirement to provide in the framework of philosophy of science such a clarification was delineated by B. Tuchańska by means of its comparison with Hempel's approach to scientific explanation as follows (1992, 105):

> Hempelian models of explanation are ... limited to the application of knowledge. To explain means to apply universal (theoretical) knowledge for understanding (empirical) phenomena that have been recorded. Hempel assumes that laws which can be used for explanatory purposes have been elaborated before an explanation starts ... This means that the construction of explanation is separated from the creation of theoretical knowledge ... *Explanation is not part of the creation of knowledge.* What is grasped by the Hempelian models is *the logical structure of applying scientific knowledge* to understand facts (particular instantiations of phenomena).

What philosophy of science should provide in addition and *primarily*, according to Tuchańska, is (1992, 105–106, 107):

> an epistemological and methodological point of view from which explanation might be considered as a certain activity which operates on scientific knowledge ... An act of explanation consists of two crucial elements: (1) existing knowledge that is modified, and (2) an epistemic context that renders some of the modifications of knowledge the acts of explanation.

1.2 Conditions in Laws and Explanations

In my approach to conditions appearing in the explanans-laws as well as in explanans of scientific explanation I draw partially on L. Nowak's approach to scientific laws and explanations (1972; 1980). He stated this approach in the framework of a philosophy of science which focused primarily on the issue of idealization. However, as I shall attempt to show, the potential philosophical and methodological yield of his approach goes well beyond the issue of idealization; the latter will not be treated here.[3]

By taking into account the practice of empirical science like physics, for example, the structure of a scientific law, L, should have the following structure:[4]

$$L^{(k)}: (x)[Ux \;\&\; Cmod_1 x = d_1 \;\&\; Cmod_2 x = d_2 \;\&\; \ldots$$
$$Cmod_k x = d_k \rightarrow F^{(k)} x = f_k(H_1 x, H_2 x, \ldots, H_r x)] \qquad (1.5)$$

---

3 On the issue of idealizations see (Batterman 2009), (Batterman and Rice 2014), (Halas 2015), (Hindriks 2013), (Jones 2013), (McMullin 1985), (Rohwer and Rice 2013), (Wayne 2011), (Weatherson 2012) and (Weisberg 2007).
4 For the semantics of this expressions see (Nowak 1972, 536).

Here, "U" stands for the universe of discourse, namely, the delineation of the type of entities for which the law is stated, "$F^{(k)} = f_k(H_1, H_2, ..., H_r)$" expresses the functional relation between a phenomenon F and $H_1, H_2, ..., H_r$ ($r \geq 1$) which are the most significant factors for it – the so-called *principal factors*, "$Cmod_i$" ($i \geq 1$) stands for a factor which is, with respect to $H_1, H_2, ..., H_r$, of secondary relevance for F and which, once at work, modifies F – I label it *modification condition*, "d_i" stands for a zero value, or for a non-zero constant value, or for an interval of values acquired by the modification condition, while k expresses the number of idealizing assumption, finally "$F^{(k)}$" and "$L^{(k)}$" stand for a phenomenon and a scientific law under such k idealizing assumptions, respectively.

As an example of a scientific law with a structure corresponding to that of (1.5) can be viewed the law for simple pendulum. It is, from the point of view of contemporary physics, stated for a conjunction of eight idealizations – $Cmod_{1-8} = d_{1-8}$ for short – and its structure can be expressed as follows:

$$(x)(Px \ \& \ Cmod_{1-8}x = d_{1-8} \rightarrow T^{(8)}x = 2\pi\sqrt{lx/gx}) \quad (1.6)$$

The idealizations given here are as follows:

1. The force of friction at the fulcrum equals zero, that is, the decrease of acceleration due to this force equals zero as well – $Cmod_1 = d_1$
2. The whole mass of the pendulum is contained in the suspended body, that is, the acceleration of the suspension cord of the pendulum equals zero – $Cmod_2 = d_2$.
3. Non-gravitational forces are not acting, that is, acceleration due to the action of non-gravitational forces is equal to zero – $Cmod_3 = d_3$.
4. The angle of deviation of the pendulum is from the interval <0°, 3°> – $Cmod_4 = d_4$.
5. The length l of the suspension cord does not change under the influence of the force of gravity, that is, changes of acceleration of the pendulum due to changes of the length of the suspension cord are equal to zero – $Cmod_5 = d_5$.
6. The volume of the suspended body equals zero; it is a mass-point – $Cmod_5 = d_5$.
7. The movement of the pendulum takes place in a vacuum, that is, the force of friction of the environment is equal to zero and so is the decrease of acceleration of the pendulum due to this force – $Cmod_7 = d_7$.
8. Forces acting on the physical system within which the pendulum is situated are equal to zero, that is, the acceleration of the pendulum due to the action these forces is equal to zero – $Cmod_8 = d_8$.

The process of explanation based on laws whose structure corresponds to that of (1.5) is based a gradual abolishment of idealizations – deidealizations, or – in the terminology of L. Nowak – *gradual concretization* unified with a gradual modification of the functional dependence of the phenomenon F.

The modification conditions play a central the role in the process of explanation by gradual concretization. Once these conditions are at work, symbolized as "$Cmod_i \neq d_i$", they gradually modify the phenomenon F as it is stated in $L^{(k)}$, so that one derives a sequence of phenomena, in symbols "$F^{(k-1)}$",..., "$F^{(0)}$". The whole sequence of concretized laws obtained by explanation by gradual concretization is as follows:[5]

$$L^{(k-1)}: (x)[Ux \ \& \ Cmod_1 x \neq d_1 \ \& \ Cmod_{2-k} x = d_{2-k} \rightarrow$$
$$F^{(k-1)}x = f_{k-1}(H_1 x,...,H_r x, Cmod_1 x)]$$

.... (1.7)

$$L^{(0)}: (x)[Ux \ \& \ Cmod_{1-k} x \neq d_{1-k} \rightarrow F^{(0)}x = f_0(H_1 x,...,H_r x,$$
$$Cmod_1 x, ..., Cmod_k x)]$$

Based on the law of the type $L^{(0)}$ one can, finally, perform the explanation of the individual phenomenon, *Fin*, by bringing in individual conditions, Cin_{1-s}. The sequence of explanations which take their course from a law of the type $L^{(k)}$ can then be expressed as follows ("–|" and "|–" stand for gradual concretization and entailment, respectively):

$$L^{(k)} -| \ L^{(k-1)} \ ... \ -| \ L^{(0)} \ \& \ Cin_{1-s} \ |- \ Fin \quad (1.8)$$

So, for example, one of the standard explanation procedures based on the law for the simple pendulum (1.6) is to suppose that the motion of the pendulum takes place under the influence of the resistance of the medium in which the pendulum moves. In this case we have to account for the force of friction F_f acting against the force of gravity and thus abolish the idealization $Cmod_7 = d_7$. But so as for this force holds $F_f = Vdg$, where V is the volume of the suspended body and d is the density of the elements of the environment per unit of volume, we also have to abolish the idealization $Cmod_6 = d_6$. We then obtain from the law with the structure of (1.6) the following law ("*m*" stands for the mass of the suspended body, "*m'*" stands here for "*Vd*"):

---

5 Here "$Cmod_{2-k} = d_{2-k}$" stands for the conjunction "$Cmod_2 = d_2 \ \& \ ... \& \ Cmod_k = d_k$". In order not to overburden the scheme, I presuppose that all idealizations can be gradually abolished.

$$L^{(6)} : (x)(Px \& Cmod_{1-5,8}x = d_{1-5,8} \& Cmod_{6,7}x \neq d_{6,7}$$
$$\rightarrow T^{(6)}x = 2\pi \sqrt{mxlx/gx(mx - m'x)})$$ (1.9)

Then, based on this law, it is possible to compute the period of a particular pendulum fulfilling the idealizations $Cmod_{1-5,8} = d_{1-5,8}$, by substituting the actual values for the magnitudes m, l, g and m'.

The sequence of explanations as reconstructed in (1.8) makes it readily seen that it involves two interconnected, but still different types of explanation methods. One stands for a sequence of derivations of laws of types $L^{(k-1)}$, $L^{(k-2)}$, ..., $L^{(j)}$, while the other stands for a derivation of an individual phenomenon *Fin*. These two types of explanation involve a taking-into-account of two different types of conditions. The former involves a taking-into-account of the action of *modification conditions* which are stated in the framework of the explanans-laws of types $L^{(k)}$, ..., $L^{(j)}$, while the latter a taking-into-account of *individual conditions* which enter into the explanans – with respect to the law of type $L^{(j)}$ – from the outside in the sense, that they are not stated in the framework of this type of law.

The phrase "initial and boundary conditions" which is standardly used in the philosophy of science for the analysis of scientific explanation has thus to be differentiated according to the nature of the conditions involved in the respective type of explanation. In the case of explanation of law from a law one should use the term "modification condition," while in the case of explanation of an individual phenomenon that of "individual conditions."

In my reconstruction of the explanation by gradual concretization as expressed in (1.8) I took into account, for the sake of simplicity, only one sequence of gradual abolishment of idealizations, namely, $Cmod_1 \neq d_1$, $Cmod_2 \neq d_2$, ..., $Cmod_k \neq d_k$, and therefore also only *one sequence* of explanation of laws by gradual concretization. But in fact we can by means of the method of gradual concretization derive several different sequences of laws depending on which idealization will be abolished as the first one, as the second one, as the third one, and so on from the set of all idealizations. We can thus obtain the following network of scientific laws (the lower index indicates which of the k idealizations has already been abolished; "⊥" stands here for gradual concretization):[6]

---

6 I suppose here that one can abolish one idealization after the other and is thus not forced to abolish several idealizations at once as, for example, in the case, given above, of the simultaneous abolishment of the idealizations that there is no resistance of the environment and that the suspended body is a mass-point. Again, I presuppose that all idealizations can be gradually abolished.

Figure 1.1: Network of scientific law obtained by gradual concretization

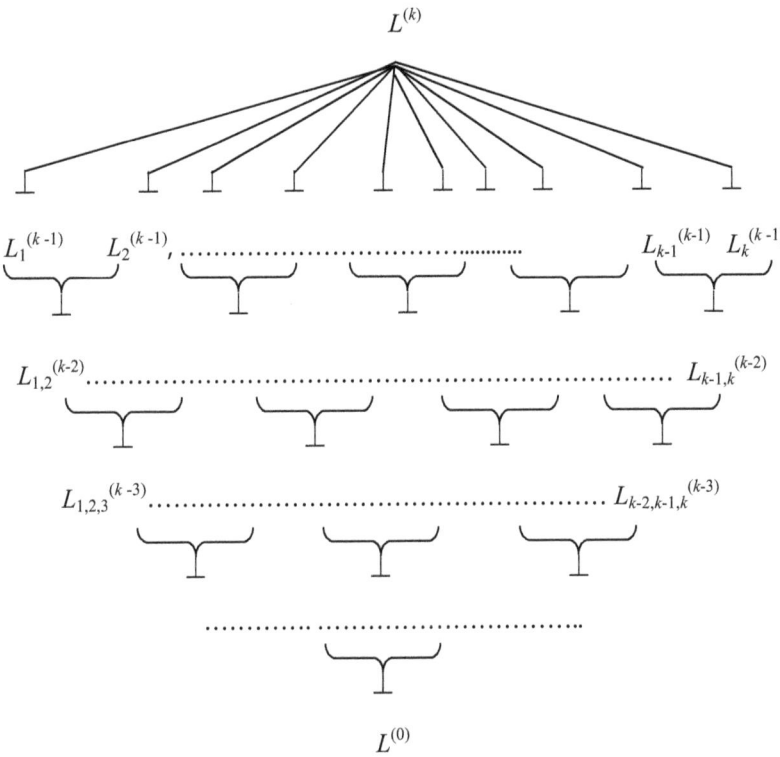

By differentiating between modification conditions which are relevant for the explanation of laws and individual conditions which are relevant for the explanation of individual phenomena it is possible to restate Woodward's (*f*)-requirement for the case of explanation of scientific laws as follows:

> (*f**) The law L_1 occurring in the explanans of a scientific explanation by gradual concretization of some explanandum-law L_2 must be stated in terms of modification conditions variations in the values of which will permit the derivation by gradual concretization of other explananda-laws $L_3,..., L_k$ which are appropriately different from L_2.

Based on the above delineated method of explanation by gradual concretization it becomes readily seen that *the more a scientific law of type $L^{(k)}$ contains modification conditions, the more different explananda-laws can potentially be derived from it by means of this method*. I emphasize *potentially* – because this method involves, as I will now try to show – in an irreducible manner several heuristic moments.

21

1.3 Heuristic in Scientific Explanation

Let us suppose that the aim of explanation is to derive an already known law. Such an explanation involves two interlinked steps. First, subsumption of this law's universe of discourse under the universe of discourse of the explanans-law of the type $L^{(k)}$ and, second, a sequence of gradual concretization which should, ultimately, yield an explanandum-law which is viewed as an explanation of the initially given law. But very often, those two universes of discourses are *prima-facie* incommensurable.

Let me take as an example of such a two-step procedure the explanation of the regularity law that bodies on an inclined plane slide down so that the distance they cover when sliding varies with the angle of inclination of the plane.

The point of departure of explanation is the second dynamic law of Newtonian mechanics which, when applied to force of gravity, involves two idealizations: (1) the body with mass m with an acceleration g due to the impact of force F of gravity has a zero volume and, (2) the body moves in a nonresistant medium. On its basis one performs a *thought-reconstruction* so that the sliding body turns into a mass-point sliding on an ideal (friction-free) inclined plane, that is, both the body and the plane turn into idealized thought-objects. Then, one states for these type of entities the equation $mdv/dt = mgsin\alpha$, where the expression on right side stands for the component of the force of gravity acting on the mass-point sliding on an ideal inclined plane with α as the angle of inclination. From this equation, together with its universe of discourse and the stated idealizations, it is then possible to derive the following law for the distance covered by a mass-point sliding on a resistance-free inclined plane:

$$L^{(4)}: (x)(Ox \ \& \ Cmod_{1,2,3,4}x = d_{1,2,3,4} \rightarrow s^{(4)}x = gxt^2 xsin\alpha x) \qquad (1.10)$$

Here "O" stands for the mass-point sliding on an inclined plane; "$Cmod_{1,2} = d_{1,2}$" stands for the conjunction of the two idealizations given already in the second law of motion, "$Cmod_3 = d_3$" stands for the idealization that the sliding mass-point starts its sliding from rest (i.e., its initial velocity is equal to zero) and "$Cmod_4 = d_4$" stands for the idealization that there is no force of friction decreasing the accelerated motion of the mass-point along the inclined plane.

Based on (1.10), one can explain, by abolishing idealization $Cmod_4 = d_4$, how the covered distance is modified once friction is at work (friction is proportional to $gcos\alpha$):

$$L^{(3)}: (x)[Ox \ \& \ Cmod_{1,2,3}x = d_{1,2,3} \ \& \ Cmod_4 x \neq d_4 \rightarrow$$
$$s^{(3)}x = gxt^2 x(sin\alpha x - cos\alpha x)] \qquad (1.11)$$

This example shows that the process of explanation unifying subsumption and gradual concretization involves non-eliminable heuristic components. The first one is given in the transformation of the universe for which the law to be explained was initially stated so that after transformation it can be viewed as falling under the universe of discourse of the explanans-law. The second is given in theoretical grasping of the impact of the respective modification condition on the relation between the respective phenomenon and the principal factors. Only once this relation is understood, it is possible to express this relation in the form of a functional relation.

What has changed also is the knowledge as expressed by the explanandum. What was, initially, known was just the regularity that bodies on an inclined plane slide down so that the distance they cover when sliding varies with variation of the angle of inclination of the plane. What was then derived was a law with the structure given in (1.11). Thus, the very explanandum underwent a profound restatement or – to use Woodward's terminology – a reconstrual in the course of its derivation by explanation. At the basis of this reconstrual is the shifting of magnitudes given initially in the explanans-law to the law already derived by gradual concretization and which were not given in the law that was stated prior to its derivation in a restated (reconstructed) form.

The transformation of the universe of discourse of the law to be explained, the gradual modification of the functional relation initially given in the explanans-law and, finally, the complete reconstrual of the explanandum-law can be viewed as a clarification of what B. Tuchańska labeled as "modification of existing knowledge" taking place in the course of scientific explanation.

The heuristic aspect of scientific explanation becomes also readily seen when in its course previously unknown modification conditions are discovered. Till now I presupposed that an explanation of a law by gradual concretization takes its course from a law in which all the modification conditions necessary for the explanation by gradual concretization are already known before this explanation is performed. The only task left here is that of performing concretizations yielding laws of the types $L^{(k-1)}$, $L^{(k-2)}$, ..., $L^{(0)}$. This task, however, can fail once the known modification conditions do not enable to perform the concretizations leading to $L^{(0)}$ which should be the explanation of a previously known law. The reason for this is that very often certain modification conditions necessary for performing the concretization procedures are initially not known. A reconstruction of that failure and of a possible the remedy to that failure can be expressed as follows.[7]

---

[7] Here I draw partially on Nowak and Nowakowa (2000).

Figure 1.2: *Failure of explanation leading to discovery of previously unknown modification conditions*

$$L^{(k)} \mapsto L^{(k+1)} \mapsto L^{(k+2)}$$
$$\bot \qquad \bot \qquad \bot$$
$$L^{(k-1)} \qquad L^{(k)} \qquad L^{(k+1)}$$
$$\cdots\cdots\cdots\cdots\cdots\cdots\cdots\cdots\cdots\cdots$$
$$\bot \qquad \bot \qquad \bot$$
$$L^{(0)}{}_\Delta \qquad L^{(1)} \qquad L^{(2)}$$
$$\qquad\qquad \bot \qquad \bot$$
$$\qquad\qquad L^{(0)}{}_\Delta \qquad L^{(1)}$$
$$\qquad\qquad\qquad\qquad \bot$$
$$\qquad\qquad\qquad\qquad L^{(0)}$$

The idea here is that once the failure (indicated as "Δ") of explanation is detected at the "end"-point of the concretization sequence, then the primary source of this failure is located in the law of type $L^{(k)}$ in the left column (the same holds for $L^{(k+1)}$ in the central column) at the top of this sequence, and it is found out that this law cannot be used for explanation of the respective law because it lacked a certain modification condition. Once the latter is found out, it is integrated (indicated as "|→") into the law of type $L^{(k)}$ ($L^{(k+1)}$) so that it turns into the law of type $L^{(k+1)}$ ($L^{(k+2)}$) from which the explanation procedure by gradual concretization can take its course anew.

Very often, the knowledge about previously unknown modification conditions is integrated into a scientific theory by drawing on other scientific theories, often called in philosophy of science as "background theories." So, for example, in the above given reconstruction of the law for the simple pendulum, the knowledge about the existence of non-gravitational forces comes from physical theories which are different from classical mechanics in which the law of pendulum was first stated. Christian Huygens in 1673 lacked knowledge about these forces; his definition of simple pendulum was as follows: "A simple pendulum is one which is understood to consist of a string, or an inflexible line devoid of gravity (*gravitas*), and of a weight (*pondus*) attached to its lowest part and whose gravity is understood to be located in one point" (1673, 92; 1986, 106).

The integration of modification conditions into a theory from other theories can have the character of a simultaneous introduction of certain idealizations and of an abolishment of certain other idealizations. This was the case, for example, in Schrödinger's derivation of the various forms of the wave equation in his *Mitteilungen* I through IV (1926a; 1926b; 1926d; 1926e).

Schrödinger stated in the first *Mitteilung* the initial form of his wave equation $\Delta\psi + \frac{2m}{K^2}(E + \frac{e^2}{r})\psi = 0$ for the hydrogen atom under two idealizing assumptions: the speed v of its electron with charge e and mass m is much smaller the speed of light c, $\frac{v}{c} = 0$, and the atom is not disturbed by external forces, $F_E = 0$.[8] This equation, once taken in unity with these idealizations, yields the following scientific law ("S" stands here for a one-electron system):

$$L^{(2)} : (x)[Sx \& F_E x = 0 \& \frac{vx}{c} = 0 \to \Delta_x \psi x + \frac{2m}{K^2}(Ex + \frac{e^2 x}{rx})\psi x = 0] \quad (1.12)$$

The scientific law with the structure (1.12) is then in the second *Mitteilung* generalized by introducing the magnitude V standing for the potential for which should hold the idealization $\frac{\partial V}{\partial t} = 0$. One then obtains a law with the following structure:

$$L^{(3)} : (x)[Sx \& F_E x = 0 \& \frac{vx}{c} = 0 \& \frac{\partial Vx}{\partial tx} = 0 \to \Delta_x \psi x + \frac{8\pi^2}{h^2}(Ex - Vx)\psi x = 0]$$

(1.13)

In the third *Mitteilung*, Schrödinger abolishes the first idealization in (1.12) and at the same time introduces magnitudes characterizing two external forces, the magnetic, with the strength F_{em}, and the electric, with the strength F_{ee}, so that the physical system is subjected only to the latter, that is, $F_{ee} \neq 0$, while the former is put equal to zero, $F_{em} = 0$, and the system under consideration is again a one-electron system. For such a system then holds the scientific law with the following structure:

$$L^{(3)} : (x)[Sx \& F_{em} x = 0 \& \frac{vx}{c} = 0 \& F_{ee} x \neq \to \Delta_x \psi x + \frac{2m}{K^2}(Ex + \frac{e^2 x}{rx} - eF_{ee} zx)\psi x = 0]$$

(1.14)

In the fourth *Mitteilung* Schrödinger departs from the law (1.13) and gives up its third idealization and derives the following scientific law:

$$L^{(2)} : (x)[Sx \& F_E x = 0 \& \frac{vx}{c} = 0 \& \frac{\partial Vx}{\partial tx} \neq 0 \to (\Delta_x - \frac{8\pi^2}{h^2} Vx)^2 \psi x + \frac{16\pi^2}{h^2} \frac{\partial^2 \psi x}{\partial tx^2} = 0]$$

(1.15)

---

8 Here "ψ" stands for the wave function, $K = h/2\pi$, where h is Planck's constant, "E" stands for energy of the one-electron system.

In a last step, he abolishes the first and second idealizations in (1.14), and so arrives at the relativistic-magnetic generalization of the wave equation and thus at a law of the type $L^{(0)}$.

Schrödinger's gradual derivation of the various forms of the wave equation taking its course from the law (1.12) thus is of a branching nature. At its basis are modification conditions which are introduced into the laws by drawing on conceptual resources of other physical theories, for example, theory of electromagnetism (magnitudes F_{ee} an F_{em}) and relativity theory (the ratio of v to c; first as $\frac{v}{c} = 0$ and then as $\frac{v}{c} \neq 0$).

1.4 Open Problems

As a conclusion, let me state some problems which follow from the above given methodological analysis of scientific explanation, and which should be solved by future research.

First, when dealing with the network of scientific laws (Figure 1.1) which can be derived by gradual concretization I presupposed, for the sake of simplicity, that the order in which the impact of the modification conditions is taken into account is of no importance for the end-result of the concretization sequence. This, however, need not hold in cases when the object under investigation displays a kind of a "memory," as for example it is the case with human actions in social systems. How to take this feature into account in the reconstruction of the method of explanation by gradual concretization remains a problem which should be targeted in the future.

Second, the above given differentiation between modification and individual conditions was based primarily on an analysis of laws and explanation procedures in physics. It remains a task of the future to complement this differentiation with its explication by means of modern possible-world semantics.

Third, when dealing with the heuristic aspect of explanation I dealt in fact just with one type of heuristic which was involved in the search for an answer to the following question: *What modification conditions were at work and which not, given a law known to be operational, so that a certain law, already known to hold, was produced?* This situation I express by the following scheme (here "????????" stands for as yet not known modification conditions):[9]

---

9   Here I draw partially on Sintonen (2005). I presuppose, for the sake of simplicity, that the law L which is subsumed in an heuristic manner under the explanans-law $L^{(k)}$ does not contain any idealizations.

Figure 1.3: Heuristic in the search for as yet not known modification conditions

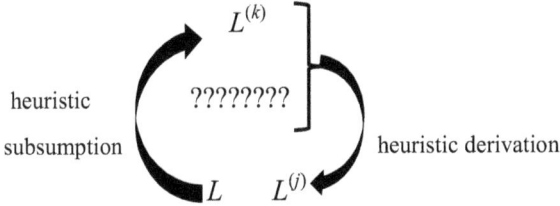

Another type of heuristic worth to be analyzed in the framework of philosophy of science is given when scientist put and search for an answer to the following type of question: *What law was (explanation) or will be (prediction) produced, given a certain law known to be at work, under the impact of certain modification conditions $Cmod_{1-j}$ known to be also at work, if other modification conditions, also known, were not (or will not be) at work, $Cmod_{k-p} = 0$?* This situation I express by the following scheme (here "*L????*" stands for an as yet not known explanandum-law which is supposed to be already produced in the past or supposed to be produced in the future):

Figure 1.4: Heuristic in the search for as yet not known explanandum-law

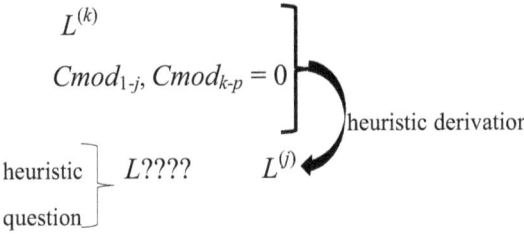

A different type of heuristic, also worth to be analyzed, obtains when the following type of question is stated by scientists: *For certain law L already known to hold, what law and modification conditions were at work under which this law was produced?* The corresponding scheme here is (here "$????L^{(?)}$" stands for an as yet not known explanans-law with an as yet unknown number of idealizations and "????????" for as yet not known modification conditions):

27

Figure 1.5: Heuristic in the search for as yet not known explanans-law and modification conditions

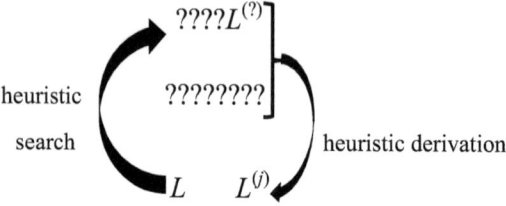

Part I:
Methodological Issues in Physics

Chapter 2: Bohr's Atom: Data, Phenomena, Laws of Phenomena, and Explanations *via* Mechanism

This chapter has several aims. The first is to unify two directions in contemporary philosophy of science: a) the direction which draws on the seminal works of J. Bogen and J. Woodward to deal with the relation of data to phenomena and b) the direction which utilizes the article (Machamer and Darden and Craver 2000) to deal with the knowledge about mechanism and its employment in scientific thinking. This unification aims to delineate the main stages of the development and growth of scientific knowledge and shall draw on the results achieved in the framework of philosophy of science in the last 20 years or so, as well as on the reconstruction of the main stages of development of knowledge leading from A.-J. Ångström's measurement of the wave-lengths of spectral lines of hydrogen in 1868 to N. Bohr's theory of the hydrogen atom proposed in 1913. In doing so, I shall draw on two sources: the history of experiments, measurement, and theory formation on the basis of original scientific articles from the field of spectroscopy and the atom of hydrogen.

The second aim, located at a meta-philosophical level, is to show that the reconstruction of the development of scientific knowledge (which is characterized in contemporary philosophy of science as a movement from data, via phenomena, to mechanisms) in fact amounts to an assignment of philosophical categories (e.g., *data, phenomena, mechanism*, etc.) to the respective stages of this development. From this meta-philosophical point of view it will come to the fore that my choice of the above given episode of history of physics can contribute to the explication of these categories, and at the same time enables, by bringing in additional categories, to broaden the network of categories by means of which philosophy of science approaches and reflects natural sciences. The categories that I shall introduce in addition to those used in the more recent philosophy of science literature (e.g., *data derived from experiments, phenomena as derived from data*, and *mechanism*) are as follows: *phenomena as derived from the law of phenomena*, and with respect to the category *mechanism* I shall also introduce the additional category pairs *ground-grounded* and *reason-reasoned*, as well as *formal ground* and *real ground*.

Finally, by analyzing certain aspects of the views of P. Machamer I shall try to show that some of the central problems of philosophy of science (e.g., the relations among data, phenomena and mechanism, the observable and the nonobservable, the empirical and the theoretical, have their origin in a misunderstanding of the nature of human thinking and of the practical operations in the world of nature.

2.1 Meta-reflections: Pure Metaphysics, Pure Epistemology, and Epistemology-*cum*-Metaphysics

In my approach I shall operate with philosophical categories assigned to the particular levels of scientific knowledge and thinking. I shall view these categories as a reconstruction of the levels of performed thinking and, via such reconstruction, as the levels of attained *knowledge about objects* or *things* (entities, processes, activities, etc.). In this sense, these categories have an *epistemological* as well as a *metaphysical* aspect, and – because they reconstruct the operations of thinking (operations in the mind, thought operations) – also a *logical* aspect. I shall characterize these operations by the category *thought operations on things in mind*. By means of the epistemological, metaphysical, and logical characterization of philosophical categories I shall try to grasp thinking as a *unity of processes* (which I shall label with the category *cognition* – in the sense of coming-to-know) *and their results* (which I shall label with the category *knowledge*).

Then, I shall gradually introduce philosophical categories corresponding to my reconstruction of sequences of thought operations and their results as given in the movement from Ångström's measurement of the wave length of the spectral lines of hydrogen in the solar spectrum up to Bohr's theory of the hydrogen atom. This introduction will follow the rule that one should assign new categories to those new determinations of thinking which arise at the respective level of scientific thinking. Stated otherwise: *my stage by stage representation of the development of scientific thinking will be such, that at no stage of representation a category will be introduced which does not correspond to a determination of scientific thinking originating in the respective level of development or which did not pass into it from the previous level; each assigned philosophical category must find its correct place.*

My epistemological/metaphysical/logical approach differs from that applied in philosophy of science dealing with categories like *data, phenomenon, mechanism*, etc. So, for example, in (Machamer and Darden and Craver 2000), one encounters the following metaphysical claim: "Mechanism occur in nested hierarchies ... lower level entities, properties, and activities are components in mechanisms that produce higher level phenomena" (Machamer and Darden and Craver 2000, 13). This is followed by an additional epistemological claim: "the descriptions of mechanisms in neurobiology and molecular biology are frequently multilevel ... Nested hierarchical descriptions of mechanisms typically bottom out in lowest levels of mechanism. These are the components that are accepted as relatively fundamental" (Machamer and Darden and Craver 2000, 13). Such an addition means that the metaphysical claim is stated from the perspective of "God's eye." Philosophy of science pretends to possess the knowledge – prior and independently of any cognition as a process – of the very ontical structure of the natural world in itself is, and then adds the epistemological claims which should describe how natural scientists discover this ontical structure of the world.

In my approach I do not need two classes of philosophical statements – metaphysical and epistemological, since the latter reconstructs the levels of the appropriation of the world of things of nature by thinking, that is, the levels of the development of knowledge about this world, rendering the former unnecessary. This "one-class" approach allows me to characterize scientific thinking epistemologically, and at the same time metaphysically and logically, thus explaining *how, by means of thought operations, scientific thinking thinks how the things of nature are.*

Accordingly, I shall consider the epistemological and metaphysical views in my approach as inseparable. This approach I shall, for the sake of brevity, label "epistemological." In my view, this solves the apparent paradox in Machamer's argument in which he claims that he is "often unsure how to do metaphysics in a way that keeps it apart from epistemic," while at the same time uniting metaphysics with epistemology: "I presume something must be said about the relations a 'thing' has to other 'things' that are in one's ontology, and specifically one must say something about the epistemic status of such things and why and how they are important or necessary for knowledge and explanation" (Machamer 2004, 28). Ultimately, I shall argue that, if one accepts my philosophical approach, it should be possible to explicate the metaphysics inherent in scientific thinking as that *which stands for a network of thought determinations into which natural scientists bring nature and make it intelligible to them.*

The application of two classes of philosophical statements, one of a metaphysical and one of an epistemological nature, the latter being seemingly based on the position of "God's eye," leads in philosophy of science to paradoxical results. For example, L. Darden thus aims at a reconstruction of the reasoning strategies used in what she labels as "the discovery of mechanism," (Darden 2002) by modeling this strategy on knowledge, *seemingly* given a priori of what a mechanism is (or, should be). Based on this approach she then declares: "Reasoning to find any particular mechanism is aided by this characterization of what a mechanism is, as well as the constraints that any adequate description of a mechanism must satisfy" (2002, 356), and at the very end of his article she declares: "The idea that what is to be discovered is a mechanism guides reasoning in its discovery. Not surprisingly, knowing the nature of the product shapes the process of discovering it" (2002, 363). So, as she is dealing here with the process of discovering of mechanism – and this process takes place in mind by means of reasoning – the term "product" should stand for "thought product" in sense of "product of thinking." But then Darden's philosophical reconstruction of the strategy of discovery ends up with the claim that, in order to perform this discovery, *one should have knowledge of what is yet to be discovered.*

It is also worth mentioning here that while Darden tries to present a seemingly *a priori* (i.e., independent of, and prior to, any knowledge as a process of cognition) philosophical knowledge of what mechanism is in itself, she does not in fact draw on a "pure" metaphysics. Instead, she appropriates this metaphysics from the knowledge given in biology, that is, from a thought product of a natural science. It is only in this way that she can build up a seemingly "pure" metaphysics of mechanism. This paradox disappears if one aims, by means of a set of logico-epistemological-*cum*-ontological categories, to reconstruct how thinking arrives at the knowledge of what mechanism is, that is, without presupposing *a priori* (i.e., from "God's vantage point) knowledge of what a mechanism is.

I shall show that the attempt at a separation of metaphysical and epistemological categories leads in the works of Bogen and Woodward to the following two problems. First, when epistemologically dealing with the movement of thinking from the knowledge about data to knowledge about phenomena, they also bring in the metaphysical issue of how phenomena cause data, but without taking into account precisely how knowledge about phenomena and data stands for knowledge about the causal (co)determination of data by phenomena. Second, even if they viewed phenomena in their epistemic status both as *derived from* data and as *explained by* theory, they did not, at the level of categorial reflections, differentiate between them in sense that these phenomena mutually differ in their respective epistemological-cum-metaphysical statuses. They stand for two different levels of knowledge of the world of nature. This difference will lead, as I shall show soon below, to a more detailed delineation of the category *phenomenon* by means of an additional categorial differentiation between two types of phenomena.

The explicit targeting of categories by philosophy of science also has the advantage of providing a criterion for a comparative evaluation of philosophies of science by means of a comparison of the richness of categorial structures which they provide, and by means of which they approach science. The importance of the recent philosophical works on data, phenomena, and mechanism lies in the fact that they enable to overcome the impoverished network of categories of philosophy of science based on the tradition of logical empiricism, which relied on two central categories: *observable phenomena* and *empirical laws* which should refer to the observable state of affairs.[10]

2.2 J. Bogen and J. Woodward on Data and Phenomena

Bogen's and Woodward's criticism targets "those who hold that scientific theories explain what we observe and who then go on to tie the relevant notion of observation rather closely to sensory perception" (Bogen – Woodward 1988, 306). In opposition to those who approach scientific theories and explanations in this way, they propose a differentiation between the category data and the category phenomena as follows (1988, 305–306):

Data ... play the role of evidence for the existence of phenomena, for the most part [they] can be straightforwardly observed. However, data typically cannot be predicted or systematically explained by theory	By contrast, well developed scientific theories do predict and explain facts about phenomena. Phenomena are detected through the use of data, but in most cases are not observable in any interesting sense of the term[11] ... Facts about phenomena ... are evidence for the high-level general theories by which they are explained.

So, in their view, while claims about data serve as evidence for claims (facts) about phenomena, claims about phenomena serve as evidence only for claims which are already part of general theories explaining/predicting claims about phenomena but not about data. Stated in a more general way: "we need to distinguish what theories explain (phenomena or facts about phenomena) from what is uncontroversially observable (data)" (1988, 314). Data can also be subjected to explanations, but because they are produced by irregular coincidences of myriad particular causes, such explanations (1988, 326):

> when they can be given at all, will be highly complex and closely tied to the details of particular experimental arrangements ... Thus, explanations of data will often lack generality ... Moreover, the factors involved in the production of any given bit of data may be so disparate and so numerous, and their co-occurrence so rare, that the details of their interaction may be both epistemically inaccessible and difficult to model theoretically.

---

10 For a classical statement of this see (Hempel 1958).
11 This claim was recently restated as follows: "phenomena need not be observable ... asking whether phenomena are observable is often not the right question to understand how ... reasoning works. This is because the reliability of ... reasoning often has little to do with how human perception works" (Woodward 2011, 171).

Exhibitions of dependency-relations of the sort that would be achieved by explicit derivations or the tracing of specific causal mechanisms may prove impossible because of computational intractabilities.

For the relation of data to phenomena it holds that the latter are derived from the former; (Woodward 1989, 396–397):

> The problem of detecting a phenomenon is the problem of identifying a signal in a sea of noise, of identifying a relatively stable and invariant pattern of some simplicity and generality with the recurrent features – a pattern which is not just an artifact of the particular detection technique we employ or the local environment in which we operate.

Bogen and Woodward explain, as already stated above, the difference between the category *phenomena* and the category *data* in terms of the category *observation*. While the former need not be observable, the latter can usually be directly observed in the following sense; (Bogen – Woodward 1992, 593):

> Data are *records* or *reports*—accessible to the human perceptual system and available for public inspection. Some data (e.g., reports of measurement results written in laboratory notebooks, or the drawings of the field geologists) are produced by human perceivers. But ... many data are produced by nonhuman measurement and recording devices ... Data constitute observational evidence to investigate phenomena.

So, for example, they claim that when measuring experimentally the temperature at which a sample of lead melts, the value – the datum – of each thermometer reading is observable (Bogen – Woodward 1988, 319).

At the same time, Bogen and Woodward give the following additional categorial characterization of the relation between data and phenomena (Bogen – Woodward 1992, 593–594). In order to detect phenomena, the experimenter triggers processes *producing data*. The crucial point here is that the produced data are reliable; this is achieved by the experimental set up aiming at, e.g., the elimination or at least the control of confounding causes. In a subsequent step, *data are interpreted*, based on a thought movement from data to conclusions, the latter pertaining to the phenomenon under investigation. Here techniques like data analysis, data reduction and data discarding are employed. With respect to both data production and data interpretation, perception figures as follows (Bogen and Woodward 1992, 596–597):

> There are two different points at which human perceivers can figure in phenomena detection. Sometimes the perceiver herself will be part of the causal chain or process leading to the *production* of data. Here what the perceiver outputs will be either data or an intermediate step in its production ... Second, whether or not human perceivers figure in data production, the *interpretation* of data (its use to support claims about phenomena) will virtually always involve human perception.

By differentiating between the production and interpretation of data, one can also differentiate between two different types of possible changes (Woodward 2000a, S165):

> While changes in data production will involve changes in the physical processes that cause the data and very often changes in the physical characteristics of the data themselves, changes in data interpretation will leave these processes and characteristics unaltered and will instead involve changes in the assumptions and patterns of reasoning that investigators bring to the data after it has been produced.

This differentiation thus means that Bogen's and Woodward's reasoning about the category *data* runs along two clearly separated lines: the *metaphysical* and the *epistemological*. In the metaphysical line data are *causally* produced in the world of nature (Woodward 1989, 393). At the same time they should be "records of effects in investigator's sensory system or experimental equipment" (Bogen 2011, 8) and also "public records ... produced by measurement and experiment" (Woodward 2000a, S163). In the *epistemological line* data are subjected to transformations *in mind* in the process of their thought interpretation. These transformations have "to do with inferring claims about phenomena from claims about data" (Woodward 1989, 393). Stated otherwise, in the approach of Bogen and Woodward, the philosophical category *data* has two clearly separated meanings: *the metaphysical and the epistemological*.

The same two lines of reasoning are also applied to, and thus also two meanings are assigned, to the category *phenomenon*. From the point of view of metaphysics "[i]t should be clear that we think of particular phenomena as in the world, as belonging to the natural order itself and not just to the way we talk or conceptualize that order" (Bogen – Woodward 1988, 321), and "[p]henomena ... are relatively stable and general features of the world" (Woodward 1989, 393) and, when compared to the experimentally produced data, are "more widespread and less idiosyncratic, less closely tied to the details of a particular device or procedure (Woodward 1989, 395).

From the point of view of their epistemology it should hold: "Phenomena ... are potential objects of prediction and systematic explanation by general theories and which can serve as evidence for such theories" (Woodward 2000, S163). This quote *I* interpret, contrary to their above given metaphysical delineation of the category "phenomena," as follows: explanations and predictions are thought operations performed on phenomena as objects or things of *mind*.

Finally, let me mention one specific feature of Bogen's and Woodward's metaphysics of data and phenomena, namely, the view that *phenomena causally produce data* in the sense that certain features possessed by data have their origin in the causal impact of phenomena and, at the same time, in the co-influence of other causes, which from the point of view of the interests of scientists, are of a secondary nature. So, for example, if data (Woodward 2000, S163–S164, S167):

> serve as evidence for the existence of phenomena for their possession of certain features [then] ... they reflect the causal influence of the phenomena for which they are evidence but they also reflect the operation of local and idiosyncratic features of the measurement devices and experimental designs that produce them ... in typical cases data are the results of many causal factors and at most some of these will have to do with the phenomenon of interest.

As an example of this, Woodward mentions the case of the measurement of the melting point of lead (Woodward 2011, 166–167):

> Schematically, if we think of the melting point as a fixed quantity M, then even if the measurement procedure is working properly the data d_i observed on different occasions of measurement will be some function f of M and of additional causal factors μ_i which will take different values on different occasions of measurement: $d_i = f(M, \mu_i)$ ... The researcher will be able to observe the measurement results d_i but will usually not be able to measure or observe the values of all of the additional causal factors μ_i. Here goal will nonetheless be to extract from the data d_i, an estimate of the value of M.

Thus, in Woodward's view, M as the "true" temperature at which lead melts is the *principal cause* and $\mu_1, \mu_2, ..., \mu_k$ are, with respect to it, *secondary causes* of the actually measured temperatures $d_1, d_2, ..., d_k$.

S. Schindler gives, without taking into account Bogen's and Woodward's differentiation between the ontological and epistemological meanings of the categories *data* and *phenomena*, the following schematic summary of their views (Schindler 2007, 167):

Figure 2.1: S. Schindler on Bogen's and Woodward's views on data, phenomena and theory

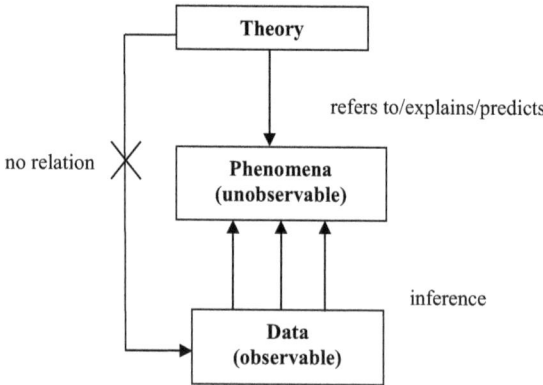

The views of Bogen and Woodward, once this differentiation is taken into account, are summarized in Table 2.1 and Table 2.2. Since they characterize the process of systematic explanation to which claims about phenomena are subjected as "explanation of an outcome due to the assertions of the details of a mechanism" (Bogen – Woodward 1988, 323), I integrate into the first table the metaphysical category *causal mechanism* and into the second one the epistemological category *theory of causal mechanism*. And, since I was not able to discern from Bogen's and Woodward's works if they hold to the view that theories about causal mechanisms are derived from claims (facts) about phenomena, I put question marks in the bottom cell of the second column in Table 2.2.[12]

Table 2.1: Reconstruction of Bogen's and Woodward's views on categories of metaphysics

Category \ Category	Data	Phenomenon	Causal mechanism
Data	are produced in experiments	causally produces	x
Phenomenon	x	x	causally produces

Table 2.2: Reconstruction of Bogen's and Woodward's views on categories of epistemology

Category \ Category	Claims about data	Claims (facts) about phenomenon	Theory for causal mechanism
Data	refer to & are interpretations of	x	x
Phenomenon	x	refer to	explains how is causally produced
Causal mechanism	x		refers to
Claims about data	x	x	x
Claims (facts) about phenomena	are evidence for/enable to derive	x	systematically explains
Theory for causal mechanism	x	?enable to derive?	x

---

12 In Schindler's figure arrows leading from phenomena to theory are not given.

2.3 From the Spectra of Gases to the Internal Motion in the Molecules of Gases and "Back": The "Stoney" Program

The beginning of spectroscopy dates back to the early 19th century and was related to the discovery of dark lines in the spectrum of the solar light as well as to experiments trying to provide information about chemical substances found on Earth by their experimental treatment by means of electricity (so-called "electrical spectra of metals") and by means of flames (so-called "flame spectra").[13] Soon afterwards, attempts were made to explain these spectra either on the basis of the corpuscular or wave theory of light,[14] then being followed by attempts to bring in the kinetic theory of matter in order to link spectra to certain chemical substances identified on Earth, which, in turn, paved the way for attempts to link the descriptions of spectra to theories of the internal structures of substances.

From the point of view of this chapter, as important appear the works of J. G. Stoney who in a series of articles (1868; 1871; 1872) delineated a general framework in which the treatment of spectra went on for several decades. In (1868) he sets as his aim to relate the characteristics of the spectra – now already understood by him as properties of the motion of light – with the hypothesized motion given inside gases, the choice of gases as the prime object of analysis being justified by the fact that at that time only their molecular composition was thought to be known with a sufficient degree of precision. There were also three additional presuppositions on which his reasoning was based. First, that "many facts in physics and chemistry lead irresistibly to the conclusion that the molecules are resolvable into simpler elements; and the probability distinctly is, that each in most gases is a highly complex system" (Stoney 1868, 133). Second, the lines of motions within these molecules of gases are or resemble orbits. And, third, light and thus also the spectral lines of gases as vibrations of a luminiferous aether filling the space between the constituents of molecules.

Stoney's reasoning can be reconstructed as consisting of two distinct but closely interrelated phases. In *phase one* he moves *from* the already given knowledge about the spectra of gases *to* the conceptual treatment of motion inside a gas. He declares (1868, 134–135, 137):

> The molecular motions consist of two very distinct parts—the motions of molecules among one another, and the motion in the interior of each molecule ... the former ... motions are irregular, and on this account ... they are unfitted to absorb or develop vibrations in the luminous æther ... But each molecule is also a separate system, the constituent

---

13 On this see (McGucken 1969).
14 On these theories see (Darrigol 2012).

parts of which are in energetic motion among themselves ... we must presume that it is they which influence the æther, producing those bright lines which constitute nearly the whole spectrum of an incandescent gas, and are in corresponding degree influenced by the æther, absorbing the same rays. How wonderfully regular the internal motions of the molecules are, and at the same time how complex, appear to be revealed to us by the fixity of the rays in the spectrum of each gas, and by their number within the limits of the visible spectrum and the doubtless greater number that exist beyond. ... If the irregularity [due to the perturbations inside the molecule] fade away so readily as to last but a trifling part of the interval between the collisions [between the molecules], the line in the spectrum will be sharply defined ... It has been observed, too, that rays of high refrangibility are more easily dilated into spectral bands of the class with which we are here dealing than rays of low refrangibility. Perhaps this may arise from the circumstance that a ray of high refrangibility is due to an atom of the molecule the periodic time of whose motion is shorter than those of other motions which give rise to rays of low refrangibility ... When we compare different gases, we find that their molecules differ both in mass and in the motions that prevail in them. That the internal motions differ is abundantly testified by the amazing variety in the grouping of spectral lines.

Finally, he provides a quantitative treatment of time a molecule moves freely between two consecutive collisions with other molecules, and based on it draws the conclusion that (1868, 140)

> It is now no longer surprising that the temporary perturbation caused while two molecules are whirling past each other, has, in many instances, had abundant time to pass away early in the interval between to collision, so as to leave the greater part of the minute interval motions [within the molecule] to be executed in the undisturbed manner which the definiteness of the spectral lines attests to us.

While in phase one, Stoney moved from the knowledge of the spectra of gases to the very gases, which he decomposed in *mind* initially to the constituent molecules, finally to arrive at the motion of an atom inside a singular molecule, in phase two, he conceptually deals with the very periodic motion within a gas in order to understand the origin of line spectra (i.e., the knowledge about the periodic motion within a gas should enable to explain the properties of its line spectra). Stated in the language of categories, the former should serve as a *ground*, in the sense of a basis for the explanation, of the latter. Stoney's statement about the ground here is as follows: "Each such periodic motion in the gas will throw off waves in the æther" (Stoney 1871, 41).

This idea is given by him an exact mathematical treatment based on Fourier's theorem, so that while the disturbance inside the gas – as the *cause* of the various spectral lines – has the period T, the effects of this cause – the spectral lines themselves – will have periods which are submultiples of T, that is, $\frac{T}{2}, \frac{T}{3}$, etc.

Based on such an understanding of the relation between the causal mechanism and the effects it produces, Stoney is able to give the actual value of the period T for three spectral lines of hydrogen, drawing on the results of measurement of their wave lengths accomplished by Ångström in 1868 with which I shall deal in detail below. The three waves length for these lines – in Frauenhofer notation C, F, and h – once reduced to vacuum, are $\lambda_C = 6563.93 \cdot 10^{-10}$ m, $\lambda_F = 4862.11 \cdot 10^{-10}$ m, and $\lambda_h = 4102.37 \cdot 10^{-10}$ m, and Stoney arrives at the conclusion that these three wave length "are the 20th, 27th, and 32nd harmonics of a wave length (*in vacuo*) of 1.3127714 fifth-metre (the Vth-metre being $\frac{1}{10^5}$ of a metre)" (Stoney 1871, 43), that is, $\lambda_C = \frac{\lambda_0}{20}$, $\lambda_F = \frac{\lambda_0}{27}$, and $\lambda_h = \frac{\lambda_0}{32}$, where $\lambda_0 = 1.3127714 \cdot 10^{-5}$ m. Based on the equation $T_0 = \frac{\lambda_0}{v}$, where v stands for the velocity of light (for Stoney v = 298·10^6 ms^{-1}), he computes the period of motion inside the molecule of hydrogen in which its spectra should have their origin: $T_0 = 4.4 \cdot 10^{-14}$ seconds.

Based on this reasoning, Stoney then accomplished, from the point of view of the future development, an important conceptual shift in the article (1872). He introduced a new scale that utilizes reciprocals of wave lengths. And as he drew on Ångström's data give in the range of 10^{-7} mm, these reciprocals were understood by him as the number of waves given in one millimeter of length and symbolized as k.

What *epistemological conclusions* can be drawn from Stoney's articles (1868; 1871)? *First*, from the point of view of thinking, his operations in the mind went from the gas to its internal structure: the molecules and their relations, to the singular molecule, then the internal structure of the molecule, finally arriving at the atom as the entity which he views as (somehow) causally relevant for the production of the gas' line spectrum. This orientation of the search for the mechanism producing line spectra of gases found its continuation in the development of spectral analysis up to the first decade of the 20th century. So, for example, J. R. Rydberg declares two decades after Stoney that "none of the physical phenomena brings us so close to the extreme limits like the spectrum of light, which allows us to study directly the movements of the very atoms" (1890a, 3). And with respect to the aim of his own work he states (1890a, 8):

> The common aim of all these researches is to arrive by means of testing and comparison of physical and chemical properties of the elements at a more exact knowledge of the nature and constitution of atoms. Today, science has generally accepted the hypothesis that all physical phenomena are owed to movements of matter. This could lead to the view that physics could one day, like chemistry, become a mechanics of atoms.

W. Ritz, nearly two decades later after Rydberg, with respect to his own attempt to give a theory of the mechanism producing spectral lines, declares (1908a, 672–673)

I do not believe that I had to suppress these deliberations even if they are not reliable. In a region, in which all "Ansätze" are missing, even such an analogy can have a certain value ... [So as] a complete theory of line spectra does not appear to be possible without special suppositions about the construction of the atoms, from the epistemological point of view it seems to be an advantage rather than a disadvantage of this hypothesis that it does not have to suppose anything about the elements from which the atoms are built ... By the way, in a question of this kind, in which a necessary inference from the effect to the cause seems barely possible, ... we barely can expect that a connection to our usual representations [*Vorstellungskreise*] would be possible *in all aspects*.

Second, also from the point of view of thinking, Stoney proposes a program aiming at a unification of two thought movements: the movement from the knowledge of the wave lengths of spectral lines of a gas to the knowledge of its central component, the atom; and *then* trying to develop conceptually, based on the knowledge about the motion of the atom, the characteristics of the line spectra. The first part of this thought movement is given primarily in his article (1868), while the intention to give the second part is present in the article (1871).

Third, a closer look at the realization of Stoney's overall project indicates the following feature. The basic period T_0 for hydrogen was derived by reasoning from the basic wave length λ_0, the latter being derived, in turn, from the wave lengths λ_C, λ_F, and λ_h of the lines C, F, and h, respectively. This *reasoning* I express by means of the categories *reason* and *reasoned*, that is, the *derived* in mind: here λ_0, and thus also T_0, are the *reasoned*, while λ_C, λ_F, and λ_h are here the *reasons*.

Fourth, λ_0, after being initially derived as the reasoned from λ_C, λ_F, and λ_h as the reasons, it should somehow be derived from the description of the internal motions of the gas. Expressed in term of categories this means that Stoney should identify in mind a mechanism, where the description of its working would be the *reason* for λ_0 as the *reasoned*. Only after such a second derivation, could the whole *cyclical* movement from the knowledge about spectral lines "back" to them be really accomplished. But Stoney does not provide the derivation which should make up the second part of this cycle. What he really obtains in his derivations is that T_0 is just the "parameter" common to the spectral lines C, F, and h, that is, he moves in his mind only from the spectral lines to the inner structure of the gas, to the atom.

Fifth, based on the above given three formulas for the hydrogen lines C, F, and h, one can state the following scientific phenomenal law unifying these formulas:

$$\forall x[H(x) \,\&\, I(x) \to \lambda(x) = \frac{\lambda_0}{s}] \tag{2.1}$$

Here "$H(x)$" stands "x is hydrogen," "$I(x)$" for "x is incandesced," "$\lambda(x)$" for "the wave length of radiation emitted by x," and "s" stands for a variable acquiring the values 20, 27, and 32. This law has the following two features. It can be used already for the purpose of explanation of the already known three different wave lengths of the spectral lines of hydrogen. It thus fulfills the requirement of the so-called *functional interdependence* imposed on scientific explanation by J. Woodward; and dealt with in Chapter 1.

Thus, I can state here the first epistemological delineation of the category *scientific law of phenomena as follows: it unifies in mind a number of different, already known phenomena pertaining to entities of the same kind and provides a unified account of these phenomena.*

The law (2.1), however, can be used neither for the computation of the wave length of fourth line known already in 1868, located before the line G, nor for the prediction of the wave lengths of spectral lines of hydrogen which could be discovered in the future.

Sixth, this means that a thorough accomplishment of the "Stoney program" has to pass through a cyclical sequence which can be characterized by the following chain of philosophical categories:

phenomena → *laws of phenomena* → *mechanism of the ground* →
laws of phenomena → *phenomena*

This scheme, however, does not take into account that Stoney in his article (1871) used in the case of hydrogen the results of Ångström, who measured the wave length of spectral lines by drawing on a set of data and on a set of experimental operations.

2.4 The Spectra and the Stability of the Atom

2.4.1 From data to phenomena: A.-J. Ångström

I now start to deal with a level of scientific knowledge to which A. Sommerfeld assigned, in a backward glance, the category *empirical data* (1923, 68), expressed in German as *empirisches* (1921, 222).

Ångström's aim was to provide a detailed atlas of the wave lengths of spectral lines given in solar light.[15] In his experiments he used two ready ready-made gratings, numbered (I) and (II), the former with 4501 lines, the latter with 2701 lines engraved on a length, declared by the producer of the gratings as being equal to 9 Paris lines. The basic equation used by Ångström for the computation

---

15 For a detailed analysis of his experiments and the conceptual basis of his computations see (Landauer 1898, 6–33) and (Baly 1912, 12–29).

of the wave length λ of the particular lines was $\lambda = e \cdot \sin\varphi$ for the spectrum of the first order, for its N-th order holds $\lambda = N \cdot e \cdot \sin\varphi$, where e stands for the width of the grating space and φ for angle of diffraction of the light waves, the latter being measured by a theodolite.

The largest part of his effort to produce the atlas consisted of the attempt to determine as precisely as possible the value of e for the gratings (I) and (II) (Ångström 1868, 4–18). This effort stood for a unity of experimental operations and of a sequence of computations. Initially, Ångström employed a dividing machine (*la machine à diviser*) usually used to produce a grating by cutting lines on a plate, made of, e.g., glass or glass covered by a layer of silver.[16] The cutting was performed by a diamond which was gradually shifted along the length of the plate; the very shifting produced by connecting a moving wheel to a screw. By a fine division of the scale on the wheel and of the screw's threads it was possible to produce gratings with values of e in the range of up to 10^{-6} mm.

Thus, since he used ready-made gratings, he employed the dividing machine for the measurement of the values of e on them. Here as central appeared to him to check the length of the screw employed in the machine. When measuring this distance between a chosen pair of neighboring lines on the gratings, he found out that this measurement yielded different results depending on which part of the screw was employed in this measurement. The most stable results were achieved by using the "double decimeter" on the screw which was found between the marks labeling this distance as being between the 400th and 600th millimeter. Ångström therefore used this section on the screw in the ensuing experimental operations and computational measurement.

In order to check if the declared and marked length of 200 mm of the "double-decimeter" on the screw was in agreement with 200 mm on the standard Paris-meter, he employed a copy of this meter produced by L.-G. Perraux. He compared the double-decimeter on the dividing machine with this meter at a temperature of 14 °C and found out that 200 mm on the Perraux-meter was equal to 200.0021 mm on the screw's double-decimeter. Ångström, however, doubted that the very Perraux-meter was a precise enough copy of the standard Paris-meter. Therefore, in a next step, he used the so-called Prussian etalon with the length of approximately 3 Prussian feet. By comparing this etalon with Perraux-meter he found out that the latter was, at a temperature of 15 °C, longer by 0.175 mm than the Paris-meter. Ångström's drive for precision, however, did not stop here. He

---

16 For the details of the construction of this machine see the url-address given for [machine à diviser] in the references at the end of this book.

employed yet another two etalon-meters, the Uppsala-meter – as the basis for the measurement of the magnitude e – and the Brunner-meter, which he regarded as the most precise copy of the standard Paris-meter, so that the latter was used by him in order to check the precision of the former. The result of this check was that at 0 °C, 1m on Uppsala-meter was 0.99810 m on the Paris-meter, with the average coefficient of thermal expansion for the former being 0.00187 for temperatures in the range 0 °C to 100 °C.

Ångström took into account also the coefficient of thermal expansion of the material from which the screw was produced, and arrived – finally – at the conclusion that the length of the double-decimeter of the screw in the dividing machine was equal to 200.0289 mm, which yielded the corrected overall length of the two gratings as (I) = 20.31717 mm and (II) = 20, 31557 mm. Then, Ångström turned to the equation for the wave length $\lambda = N \cdot e \cdot \sin\varphi$ given above, and modified it in such a way that it took into account the deviations of temperature t from temperature $t_0 = 16$ °C, as well the movement of the grating with the respect to the annual the movement of the system composed of the Sun and the Earth. At the same time, he abstracted from the influence of grating's movement around its own axis, because its speed was negligible when compared with the speed of light, as well as from the deviations of atmospheric pressure from its normal value (760 mm of height of the mercury column), because this variation had no sensible influence on the value of λ (1868, 14, 19). By this modification he obtained the relation $\log \lambda = \log \sin\varphi + \log \frac{e}{N} + \zeta$, where ζ stands for corrections brought in with respect to the orientation of the grating and the variation of temperature.

Based on this equation he then computed, based on the measured values of e, φ, and ζ for gratings (I), the data for the spectra C, F of hydrogen; the values were as follows (1868, 18):

Table 2.3: Data for the spectral lines C and F of hydrogen computed in (Ångström 1868)

Line	1. spectrum in 10^{-7} mm	Number of observations	2. spectrum in 10^{-7} mm	Number of observations	3. spectrum in 10^{-7} mm	Number of observations
C	6562.27	2	6561.97	3	6562.16	1
F	4860.67	2	4860.62	2	4860.80	3

And then, based on all the data for grating (I), as well as all the data for grating (II), he computed the *weighted arithmetic mean* values of the wave lengths of the spectral lines of solar light. For the four lines C, F, near G, and h – in the more modern notation H_α, H_β, H_γ, and H_δ to which I shall hold from now on – of hydrogen he obtained the following mean values (1868, 31–32):

Table 2.4: Average wave lengths of the spectral lines of hydrogen computed in (Ångström 1868)

Line	Wave length in 10^{-7} mm
H_α	6562.10
H_β	4860.74
H_γ	4340.10
H_δ	4101.20

These values were in 1884 used by J. J. Balmer in his formulation of the first truly unifying formula for the wave lengths of the spectral lines of hydrogen.

What *epistemological lessons* can be drawn from Ångström's experimental as well as thought operations? First, Ångström's *experimental* operations can be characterized by means of categories introduced by Bogen and Woodward (Bogen and Woodward 1988, 324–329; Woodward 1989, 411–420), namely, *practical control of possible confounding factors* as well as *replication* and *calibration*. *Practical control of possible confounding factors* stands for the cases when Ångström experimentally took into account factors affecting the outcomes of the experiments, like, e.g., the change of temperature. *Replication* stood for such an aspect of his experimental operations that he tried to set them up in such a way that they could be reproduced in other laboratories. This, in turn, required a *calibration* of the instruments employed in these laboratories, which would conform as closely as possible to one, universally accepted set of measurement units, say, the standard Paris-meter.

Second, in his *thought* operations, Ångström employed statistical analysis based on the weighted statistical mean, which enabled him to pass in mind from what I label by the philosophical category *data* to what I label by the philosophical category *phenomena*. Thus, *data* as a philosophical category stand in my approach for an *entity in mind* and simultaneously for *a certain level of attained knowledge about the world*. Both are information about occurrences given to us in our practical operations (manipulations) with things in the world; *occurrences in the world of things are thought in mind as data*.

In the case of the above given Table 2.3, the data for spectra stand for the meaning of a sentence like "The wave length of the line C (H_α) in the 1st spectrum on grating (I) is 6562·10^{-7} mm." In my view, contrary to that of Bogen and Woodward, the meaning of sentences of this type does not refer to something observable in the world, or, stated in more general categories, to something accessible to human sensory perception. *Occurrences as states of affairs in the world in the range of 10^{-7} mm are not perceptible* and, of course, the *very meaning, in the sense of information, stated in a sentence like the given above, is not perceptible either.* So, in my reconstruction here *data do not refer to perceptible state of affairs,* and *so neither do phenomena derived from them.* In fact, the meaning of a statement is never perceptible, since meanings as information can only be thought because they are entities in the mind on which one can perform thought operation which yield other thought entities. In Ångström's approach such operations where, e.g., the computations of the weighted arithmetic mean for the wave length and his abstraction from the impact of changes of the atmospheric pressure.

My view with respect to the epistemological characterization of thought operations mediating between data and phenomena differs from that of Woodward. He characterizes the thought movement leading from – what he labels as – data claims stating the particular temperatures of the melting of lead to the computed mean temperature as follows (2011, 167):

> When a thermometer is used to measure the melting point of lead, the researcher certainly hopes that the data obtained will reflect the causal influence of the temperature of the sample as it melts, along with various other factors. In this sense, the melting point M will figure in a singular causal explanation of data.

As shown above, he uses the notation $d_i = f(M, \mu_i)$, where d_i stands for the data, according to him observed on different occasions, M for principal cause and μ_i for the secondary causes being at work. Based on my exposition of Ångström's thought movements, I propose the following reconstruction of the movement from data to phenomena. Once data are given, they are subjected to a statistical treatment in order to eliminate from them in mind the *effects of the secondary causes,* with the aim to obtain the *effect of the principle cause,* and where the principal cause is not cognized, while the secondary causes are either not of primary interest for the scientists or, like the principal one, also not cognized at this level of knowledge and thought operations. Thus, from the point of knowledge, the particular data, symbolically represented by Woodward as d_i, stand in my approach for the – in mind – *supposed common effect of the combined action of the principal cause and of the secondary causes, where former, and sometimes also the latter will be understood in the future.*

In my approach, the category *phenomenon* characterizes the knowledge, derived from the knowledge characterized by me via the category *data,* of what is supposed to be the effect of the principal cause, the latter as yet to be discovered. The phenomena are thus the starting point of another set of thought operations aiming to arrive at the understanding of this cause. Bogen's interpretation of the relation, expressed as $d_i = f(M, \mu_i)$, namely, that in order to identify a phenomenon, one has to make inferences from a data set to the causes in this set, expresses one possible epistemic situation. But, the situation I propose shows up as the more simple one: *by reducing in mind the data to a phenomenon, one at the same time should obtain the knowledge about the principal cause of these data.*

In my approach I provide an epistemological reconstruction of an epistemically more complex case, whose reconstruction will yield a set of philosophical categories of which Bogen's and Woodward's set of categories is only its subset, and where my set reconstructs the thought movement as a movement *from data to a phenomenon*, and *from the latter, in the ensuing steps, to its principal cause*. Worth to be noted here is also that Woodward's example of the measurement of the melting temperature of lead is does not correspond to what he claims it illustrates. The mean, computed temperature of this melting is not the principal cause of the particular melting points measured on the basis of actually performed experiments, because no temperature is the cause of other temperatures. So, for example, even in phenomenological thermodynamics, temperature is not understood as the cause of the states of a gas. The variable labeled "temperature" is just assigned to these states as a magnitude characterizing these states.

If one looks thoroughly at the operations described in (Ångström 1868), it becomes readily seen that what is *central* to the whole endeavor here is not the difference between what is and what is not observable or, stated in more general terms, between what is and what is not *perceptible*, but the interconnection of two different but still interconnected ways of human operations: the *practical operations in the laboratory* and the *thought operation in mind*.

My epistemological reflections result in the following reinterpretations of Bogen's and Woodward's category pair *data-phenomena*.

Table 2.4: Reinterpretation of Bogen's and Woodward's category pair data – phenomena

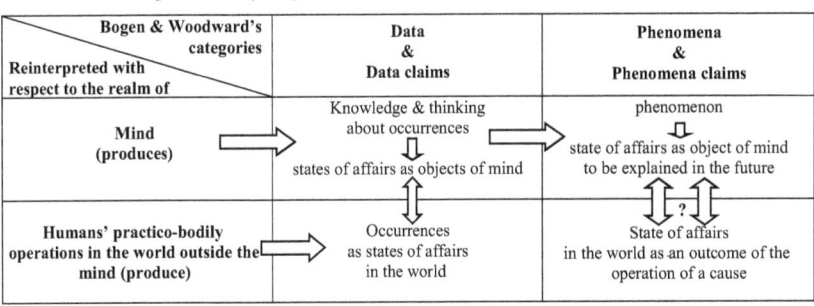

49

The horizontal one headed arrows stand for the direction of production, so that in the first line both these arrows stand for production in the mind, while in the second line it stands for material ("physical") production in the world outside the mind. The vertical one headed arrows stand for the assigning of data to *states of affairs as objects of mind* and of a phenomenon to a *state of affairs as an object of mind to be explained in the future*, where under an *object of mind* or *thing in mind* I understand an *ens rationis* in the sense which will be explicated below. The vertical double headed arrows stands for a *possible* relation of mutual correspondence between the world of mind and the world outside the mind, so that where only one such arrow is given, the experimental scientists *knows* that the particular data-claims correspond to particular states of affairs in the world outside the mind; two vertical double headed arrows, with a question mark between, express that the scientists who derived in mind the phenomenon-claim from a set of data-claims do not as yet know, if the former corresponds at all to a state of affairs (given in the world outside the mind), and if yes, the outcome of what cause it is.

2.4.2 From Phenomena to Laws of Phenomena: J. J. Balmer, J. R. Rydberg, and W. Ritz

The results given in (Ångström 1868) were the point of departure for the development of spectral analysis leading to what W. Ritz labeled by the epistemological category *empirische Gesetze* (1903, 266) and N. Bohr by the category-concept hybrid *empirical spectral laws* (Bohr 1922a, 1).

While Stoney, as shown above was able to express the knowledge of the wave lengths of three spectral lines of hydrogen, in modern notation, H_α, H_β, and H_δ, in one formula which could be used for explanation but not for prediction, J. J. Balmer was able to derive one formula unifying the knowledge of the wave lengths of the four spectral lines H_α, H_β, H_γ, and H_δ.[17] He drew on the wave-lengths of the spectral lines of hydrogen as computed by Ångström: $\lambda_\alpha = 6562.10 \cdot 10^{-7}$ mm, $\lambda_\beta = 4860.74 \cdot 10^{-7}$ mm, $\lambda_\gamma = 4340.10 \cdot 10^{-7}$ mm, and $\lambda_\delta = 4101.20 \cdot 10^{-7}$ mm. He explicitly states why he is targeting the wave lengths of hydrogen's line spectra and not of another substance (1885a, 551):

> The hydrogen, whose atomic weight is the smallest among all the hitherto known substances and characterizes it as the most simple element ... seems to me to be suitable

---

17 For an analysis of this derivation see (Banet 1966) and (Banet 1970); the latter article draws on Balmer's manuscripts.

more than any other body, to open new vistas of research about the nature of matter and of its properties. And here especially the wave length of the first four lines of hydrogen stimulate and draw the attention.

And then he sets as his aim to find: "a simple formula ... with the help of which the four outstanding lines of hydrogen could be represented" (1885a, 552). Such a formula for the wave lengths of the hydrogen lines would be an "expression of a law" (1885a, 552).

He starts by unifying the above given four wave length by means of a basic wave length $\lambda_0 = 3645.6 \cdot 10^{-7}$ mm, so that they are given as multiples of the following coefficients $\frac{9}{5}, \frac{4}{3}, \frac{25}{21}$, and $\frac{9}{8}$, respectively. By multiplying the second and fourth coefficients by $\frac{4}{4}$ he obtains four coefficients which can be written as, $\frac{9}{9-4}, \frac{16}{16-4}, \frac{25}{25-4}$, and $\frac{36}{36-4}$, and thus unified into the formula $\frac{m^2}{m^2-2^2}$, where m stands for the positive whole numbers 3, 4, 5, and 6. The four wave lengths $\lambda_\alpha, \lambda_\beta, \lambda_\gamma$, and λ_δ can be then unified into the formula

$$\lambda = \lambda_0 \frac{m^2}{m^2 - 2^2} \quad (m = 3, 4, 5, 6) \tag{2.2}$$

Formula (2.2) can be embedded into a phenomenal spectral law, so that holds:

$$\forall x \, [H(x) \, \& \, I(x) \rightarrow \lambda(x) = \lambda_0 (\frac{m^2}{m^2 - 2^2})] \tag{2.3}$$

Here the expression in the antecedent, have the same meaning as in the law reconstructed in (2.1).

The importance of such phenomenal law is that that it stood in Balmer's reasoning not only for a unification of antecedently known phenomena, but was also used by him for the derivation of phenomena, whose referents were not known to him when he stated formula (2.2). In his manuscript he wrote that "the presented formula ... would enable to determine by mere computation other still unknown lines of hydrogen" (1884, [5]). In fact, on the basis of formula (2.2) and the hypothesized value $m = 7$, he predicted the existence of another spectral line of hydrogen, with the predicted wave length $\lambda = 3969.65 \cdot 10^{-7}$ mm. Only later he learned that such a spectral line was already discovered together with other spectral lines symbolized as $H_\varepsilon, H_\zeta, H_\eta, H_\theta$, and H_ι.[18] Thus, the scope of applicability

---

18 On these spectral lines see (Von Helmholtz 1880), (Vogel 1880), and (Huggins 1880).

of formula (2.2) was extended to the values 7 through 11 of the parameter m. In addition, in the second communication (1885b) he again broadened, based on the additional knowledge he obtained about spectral lines, the scope of application of the formula (2.2), so that the values of the parameter m ranged now for the positive whole numbers 3 through 16.

Based on the initial computations leading to formula (2.2) with the coefficients, $\frac{9}{5}, \frac{4}{3}, \frac{25}{21}$ and $\frac{9}{8}$, he stated (1885a, 553)

> For various reasons it is probable for me that the above stated coefficients belong to two series, so that the second series takes once again the elements of the first series. So, I arrive at a more general representation of coefficients ($\frac{m^2}{m^2-n^2}$), where m and n are always whole numbers.

Thus, he generalizes the above given formula (2.2), so that now it turns into

$$\lambda = \lambda_0 \frac{m^2}{m^2 - n^2} \qquad (2.4)$$

The phenomenal spectral law (2.3) then turns into

$$\forall x [H(x) \& I(x) \rightarrow \lambda(x) = \lambda_0 (\frac{m^2}{m^2 - n^2})] \qquad (2.5)$$

The generalization (2.4) he stated explicitly as a hypothesis because in 1885 it was confirmed only for the value 2 of the parameter n, while (1885a, 559):

> From the lines of hydrogen which would correspond to the formula for n = 3, 4, none is found among the spectra known till now. They have to be developed probably under completely different temperatures and pressure, in order to become visible.

In the ensuing development of spectral analysis, phenomenal spectral laws were generalized in order to cover also the spectral lines of substances other than hydrogen. As crucial appear here the works of J. R. Rydberg and W. Ritz.

In (1890a)[19] Rydberg drew on Stoney's term "wave number," abbreviated now as n in the sense of "*the number of wave lengths in 1 centimeter*" (1890a, 13), that is, $n = 10^8 \cdot \lambda^{-1}$, where λ is expressed in units used by Ångström, He also drew on the already acquired knowledge that the spectra of particular substances are ordered in series. Based on this he found that the wave numbers assigned to the spectral lines of different substances can be described by the following unified formula (1890a, 40):

---

19 See also (Rydberg 1890b; 1890c; 1890d).

$$n = n_0 - \frac{N_0}{(m+\mu)^2} \quad (2.6)$$

Here "m" stands for a positive whole number labeled "number of the term;" N_0 is a universal constant, while constants n_0 and μ stand for specific constants characterizing particular spectral series of a substance, and where n_0 is the limit which n approaches when $m = \infty$. By transforming Blamer's formula (2.2) so that it now holds for wave numbers ($\lambda = \frac{10^8}{n}$ and $\lambda_0 = \frac{10^8}{n_0}$) he obtained the formula $n = n_0 \cdot \frac{m^2 - 2^2}{m^2}$ and thus $n = n_0 - \frac{4 n_0}{m^2}$. By comparing the latter formula with formula (2.6) he found out that for its universal constant holds $N_0 = 4 n_0$ and so as $\lambda_0 = \frac{10^8}{n_0}$, he computed in (1890a) that $N_0 = 109721.60$.

The development of spectral analysis showed also that substances other than hydrogen displayed often a more complex system of spectral lines, namely, doublet and triplet spectra, whose line were labeled as "principal," "sharp," and "diffuse" series.[20] The difference between the spectrum of hydrogen and those of other substances Rydberg expressed as follows (1890a, 139):

> The elementary line spectrum of hydrogen differs from the spectra of all other known elements by its exceptional simplicity. In this spectrum there are neither different groups nor double or triple lines. ... Here one finds only one simple series, and this series can be represented by the formula of Mr. Balmer, which is also more simple than the equations of all the other known series. By expressing it in the manner which we have chosen, it obtains the form
>
> $$\frac{n}{1097216} = \frac{1}{(1+1)^2} - \frac{1}{(m+1)^2}$$
>
> that is, there are no constants except N_0 and unit.

This view on the simplicity of the line spectrum of hydrogen, as compared to that of other substances, was shattered in 1896, when E. C. Pickering discovered in the spectrum of the star ζ *Puppis* a series which he interpreted as belonging to hydrogen.[21] So as it could not be described by Balmer's formula (2.2), Pickering proposed the modified formula $\lambda = 4650 \cdot \frac{m^2}{m^2 - n^2} - 1032$. Rydberg in (1897)

---

20 For the formulas for these series see (Rydberg 1896).
21 On this see (Pickering 1896a; 1896b; 1897a; 1897b).

analyzed the spectral lines discovered by Pickering and showed the Balmer series of hydrogen and Pickering's series have the same limit n_0. From this he drew the conclusion that both are spectra of hydrogen. In a next step, he characterized the Balmer series, written now as $n = n_0 - \dfrac{N_0}{(m_1+1)^2}$, as corresponding to the diffuse series of hydrogen, while for Pickering series holds $n = n_0 - \dfrac{N_0}{(m_2+\frac{1}{2})^2}$, and corresponds to the sharp series of hydrogen. Based on his approach in (1897) he combined the last two formulas and obtained the formula $n = \dfrac{N_0}{(m_1+1)^2} - \dfrac{N_0}{(m_2+\frac{1}{2})^2}$, which he, for $m_2 = 1$ and for a varying m_1, interpreted as belonging to the principal series of the hydrogen spectrum and which was not at that time observed. Rydberg, based on $N_0 = 19675.00$ and for the values of the parameter m_1 equal to 1 through 5, predicted the values for n and λ, which were later on confirmed.[22]

Finally, in early 20th century, W. Ritz formulated "a new law of the series spectra" (1908b, 521), enabling to combine the already known formulas of spectral series of a substance, so that one obtains new formulas of other series of this substance, that can be used for the prediction of the existence of these other series. This law he stated in an autoseminar paper added to his article (1908b) as follows (1911, 162):

> While the hitherto known laws of the series spectra connect with one another the lines of *one* series ... it is shown here, that there exists a simple relation also between different series of one element: by means of additive or subtractive combinations, either of the series formula or constants given in them, new formulas are being formed, which allow to compute completely the new lines discovered in recent years from those known earlier.

So, for example, as shown above, for the lines H_α, H_β, and H_γ, Balmer's formula (2.2) in Rydberg's reformulation yields:

$$n_\alpha = N_0(\frac{1}{2^2} - \frac{1}{3^2}), n_\beta = N_0(\frac{1}{2^2} - \frac{1}{4^2}), n_\gamma = N_0(\frac{1}{2^2} - \frac{1}{5^2})$$

By applying Ritz combination principle, one obtains from these three formulas the following two additional formulas:

$$n_\beta - n_\alpha = N_0(\frac{1}{3^2} - \frac{1}{4^2}), n_\gamma - n_\alpha = N_0(\frac{1}{3^2} - \frac{1}{5^2})$$

---

22 On this see (Maury and Pickering 1897) and (Lockyer and Chisholm-Batten and Peddler 1901).

Thus, one predicts here the existence of a spectral line for which should hold the formula

$$n = N_0(\frac{1}{3^2} - \frac{1}{m^2}) \, (m = 4, 5)$$

This prediction was confirmed by Paschen (1908), who – based on a series of experiments – proved the existence of a spectral series corresponding to the last formula. This discovery was at the same time a confirmation of Balmer's universal formula (2.4) for hydrogen which in Rydberg's restatement is:

$$n = N_0(\frac{1}{n^2} - \frac{1}{m^2}) \qquad (2.7)$$

while Balmer's initial formula (2.2) is now restated as

$$n = N_0(\frac{1}{2^2} - \frac{1}{m^2}) \qquad (2.8)$$

The phenomenal spectral law (2.5) then turns into the phenomenal spectral law

$$\forall x[H(x) \wedge I(x) \rightarrow n(x) = N_0(\frac{1}{n^2} - \frac{1}{m^2})] \qquad (2.9)$$

Here "$n(x)$" stands for "the wave number of the spectral line emitted by x."

What *epistemological lessons* can be drawn here? *First*, as shown above, Balmer's formula (2.2) – and thus also the law (2.3) – can be used not only (as Stoney's formula) to explain the already known wave lengths, but also to *predict* particular wave lengths. In addition, Ritz's law (2.8), like its anticipation by Balmer (2.5), can be used for the explanation of the law for Balmer's series as well as of the law for Paschen's series. I can thus provide a reinterpretation of the category *law of phenomena* given above in 2.4.2 *A scientific law of phenomena unifies in mind a number of different phenomena and laws of phenomena which pertain to entities of the same kind and provides a unified account of them.*

Second, so as the law of phenomena does not stand for the knowledge of causes, once it is used for explanations and prediction of phenomena, these two operations in the mind differ in the respective epistemic characteristics. These characteristics can be explicated by turning to Balmer's formula (2.2). The latter can be used for an "*explanation*" of those wave lengths of spectral lines of hydrogen with which Balmer was acquainted already before he derived formula (2.2), and where the knowledge about them was used by him in the derivation of this formula. I put here the category *explanation* into quotation marks because by such a derivation no extension of knowledge was obtained on his part about these particular phenomena. Using Woodward's term "reconstrual of the explanandum", no such reconstrual took place. The explananda phenomena are exactly those which were the basis for the thought derivation of formula (2.2). Stated otherwise: the cycle of thought operations mediating between knowledge characterized by the chain of philosophical categories *phenomena* → *laws of phenomena* → *already known phenomena* does not lead to an extension of knowledge.

In the case of prediction the situation is epistemically different. Based on the thought introduction of $m = 7$, Balmer was able to predict in the mind the existence of a spectral line of hydrogen about which he had no knowledge at the time when he derived formula (2.2); here his *knowledge about phenomena was extended*. This extension can be characterized by means of the following chain of philosophical categories *phenomena* → *laws of phenomena* → *previously unknown phenomena*.

What unifies both such explanations and predictions is that none of them is based on knowledge about the action of mechanism. Since law (2.3) does not express such knowledge, neither the explananda nor the predicanda derived on the basis of this law express knowledge about action by which a mechanism produces the phenomena.

2.4.3 Towards the Atom: Its Stability and Its Electrons

P. Schuster declared in his article "Genesis of Spectra" (1883, 120):

> It is the ambitious object of Spectroscopy to study vibrations of atoms and molecules in order to obtain what information we can on the nature of forces which bind them together. The vibrations we know must be of a very complicated nature, yet it is natural that not many years after Spectral Analysis was raised to the rank of science ... attempts were made to discover a law in the apparent irregularity with which different lines of the same element are distributed over the spectrum.

Thus, according to Schuster, the laws stated for spectral lines emitted by molecules and atoms should be the point of departure for the movement of cognition aiming at obtaining knowledge about the very molecules and atoms. This movement he characterizes as follows (Schuster 1883, 120):

> When different elements combine together the vibrations of the compound molecule are not obtained by the simple addition of the periods of the elements. The spectrum of a molecule is entirely distinct from that of its elements, and we may well ask the question whether we can trace in the spectrum of the compound the influence of the different atoms composing it.

And when answering this question he states his doubts as follows (Schuster 1883, 120–121):

> [W]e must not to soon expect the discovery of any grand and very general law, for the constitution of what we call a molecule is no doubt a very complicated one ... We know a great deal more about the forces which produce the vibrations of sound than about those which produce the vibrations of light. To find out the different tunes sent out by a vibrating system is a problem which may or may not be solvable in certain special cases, but it would baffle the most skillful mathematician to solve the inverse problem and to find out the shape of a bell by means of the sounds which it is able of sending out. And this is the problem which ultimately spectroscopy hopes to solve in the case of light.

Spectroscopy was, however, able to state formulas for the Balmer, Paschen and Pickering series, and even unify – by using Ritz's combination principle – the formulas for the first two series into one. Thus, as a next step, what suggested itself is the attempt to unify these general formulas with a theory of the constituent components of hydrogen gas and of atoms and molecules in general. As also shown above, the gradual development of spectroscopy led to a thought decomposition of gases into molecules, then to the targeting of a single molecule and then, in turn, to its decomposition into atoms. In the ensuing development, the discovery of the elementary quantity of electricity, labeled by J. G. Stoney as "electron" which he associated with a charge given in each atom, and whose movement he viewed as the source of spectral lines (Stoney 1891, 583; 1894), culminated in the discovery of the electron as a particle inside atoms (Thomson 1897).

Accordingly, any attempt at a unification of the laws obtained by spectral analysis with a theory of atom required to provide a thought construction of the atom so that it would take into account the movement of the constituent electrons.[23] In the first decade such a construction was provided first by J. J. Thomson and then by E. Rutherford. The former proposed a thought construct according to which electron were embedded into a sphere of positive electricity. Such a construction was sufficient to explain the mechanical stability of the atom because, as latter noted by Bohr (1912, 136; 1913a, 1–2), it added the electrons inside this sphere so that the strength of attractive forces increased with the distance from the center of the sphere. Thomson's construct failed, however, with respect to three known facts. First, once the production of the spectra is related to the motion of electrons inside the atom, it is not possible – due to the continuous nature of this motion – to explain the discrete nature of the line spectra. Second, as shown by Baron Rayleigh (Strutt 1906, 290–291), due to the *supposed* applicability of

---

23 On this see (Conway 1907) and (Bevan 1910).

classical electrodynamics to this continuous motion, the atom should continuously radiate, and thus would be instable. The validity of the laws for spectral lines suggested, however, that the atoms endure. Third, this model was not able to come to terms with results obtained by the experimental bombardment of a thin foil of matter by α and β particles.[24] This failure led Rutherford to propose his thought construction of the atom which supposed that these particles are deflected in one collision with a single atom, and where the latter "consists of a central charge and surrounded by a uniform spherical distribution of opposite sign" (Rutherford 1911, 677).[25] Even if this thought construct fared here better than Thomson's, it fared worse with respect to the issue of mechanical stability. For a certain number of electrons on a ring shaped orbit, since they exert repulsive forces on each other, the construct would collapse. And, Rutherford's atom, like Thomson's, failed to stabilize the atom from the point of view radiation. Rutherford was aware of the fact that the atom he constructed could no deal with issue of stability, but he postponed the theoretical treatment of this fact. "The question of the stability of the atom proposed need not be considered at this stage, for this obviously depend upon the minute structure of the atom, and on the motion of the constituent charged parts" (1911, 671).

The location the solution of the problems encountered by both Thomson's and Rutherford's atoms was delineated by Baron Rayleigh as follows (Strutt 1906, 291):

> An apparently formidable difficulty ... stands in the way of all theories of this character. How can an atom have the definiteness which the spectroscope demands? It would seem that variations must exist in (say) hydrogen atoms which would be fatal to the sharpness of the observed radiation; and indeed the gradual change of an atom is directly contemplated in view of the phenomenon of radioactivity. It seems an absolute necessity that the large majority of hydrogen atoms should be alike in a very high degree. Either the number undergoing changes must be very small or else the changes must be sudden, so that at any time only a few deviate from one or more definite conditions. It is possible, however, that the conditions of stability or exemption from radiation may after all demand this definiteness ... According to this view the frequencies observed in the spectrum may be not frequencies of disturbances or of oscillation in the ordinary sense at all, but rather form an essential part of the original constitution of the atom as determined by the conditions of stability.

This direction of deliberations, namely, to relate the frequencies of the spectral lines of atoms with the conditions of their own stability, was taken by Bohr. So as

---

24 On these experiments see (Geiger and Marsden 1909).
25 On this see (Heilbron 1968).

Rutherford's atom failed on the issue of stability both mechanical and radioactive, but was successful with the respect to the explanation of the deflection of α and β particles, Bohr could have taken several paths to solve the problems which beset Rutherford's thought construct. One was to mechanically stabilize it by finding out the highest possible number of electrons on a ring at which the atom would still be stable, and then in mind pass to the next outer ring, and placing there additional electrons up to the next threshold of mechanical stability. This would not, however, stabilize the atom radioactively due to the supposed applicability of classical electrodynamics to the motion of electrons in the atom. The second option was to choose such a simple type of atom in which the problem of mechanical stability would not arise due to the absence of several electrons orbiting on the same ring. But, even by means of such choice, the atom would not be radioactively stabilized, again due to the supposed applicability of classical electrodynamics. The third option, and the one actually taken by Bohr, was to unify the first and second paths and at the same time to question the applicability of classical electrodynamics to the inner structure of the atom.

This direction of thought development was taken in Part I through III of his article (Bohr 1913b) which unified in itself the results he presented in the lecture on the theory of metals in December 1911, as well as in the so-called "Rutherford Memorandum," and in the article (1913a). In the lecture he arrived at the conclusion that "the Maxwell equation for the electromagnetic phenomena are not exactly satisfied with regard to the motions of the single electrons" (1911, [419]). In the memorandum, in order to stabilize the atom, Bohr introduced the "special" non-mechanical hypothesis $E = K \cdot \nu$, that is, "that for any stable ring (any ring occurring in natural atoms) will be definite ratio between the kinetic energy of an electron in the ring and the time of rotation" (1912, [37]).[26] For this hypothesis, as he declares that "there will be given no attempt of a mechanical foundation as it seems hopeless" (1912, [37]). Finally, Bohr choose the atom of hydrogen as the point of departure for his thought operations, because he came in the article (1913a) to the conclusion that it contained only one electron orbiting the nucleus, thus, by such a choice of the thought thing he could bypass, for the time being, the conceptual treatment of the mechanical stability of his thought construct.

What *epistemological conclusions* can be drawn here? P. Schuster, as shown above, characterized the problem faced by spectral analysis of his time as that of finding a solution to what in physics is called "the inverse problem," that is,

---

26 In the equation $E = K \cdot \nu$, "ν" stands for frequency of rotation and not time of rotation, "K" stands for a constant.

expressed by means of philosophical categories, as a thought movement from the knowledge of *effects* – expressed in the knowledge of phenomena and their respective laws – to the *cause*, that is, to what I labeled above by the category *ground*, of these effects. This thought movement was accomplished by Rutherford in the thought derivation of the structure of the atom based on the knowledge of the characteristics of the scattering of particles by an atom. Rutherford expressed this epistemic situation as follows: "Since the α and β particles traverse the atom, it should be possible from the study of the deflexion to form some idea of the constitution of the atom to produce the effects observed" (1911, 670).

As shown above, while the proposed model of atom was sufficient to explain this very deflection, it was in some crucial aspect conceptually incomplete. Thus, it was necessary to turn in the mind to the atom itself and deal conceptually with its inner structure independently – for the time being – from the reflections about the spectral lines, in order to come to terms with their *cause* in the sense of their *ground*. Stated in philosophical categories, one has to distinguish two types of knowledge characterized by the category *ground*. One, when it draws only on the knowledge about the phenomena and their laws so that the ground is in mind arranged according to this knowledge. This means that the ground has, in the reasoning taking place here, the epistemic status to which the category *the reasoned* can be assigned, while the phenomena and their laws have in this reasoning the status characterized by the category *the reasons*, even if it is supposed that the *ground* causally produces, that is, *grounds* these phenomena and their laws; in the causal context the latters are the *grounded*.

One could here, seemingly, reverse in mind the relation between the knowledge of phenomena and their laws, on the one hand, and the knowledge of the ground, on the other, so that the ground would be the reason while the phenomena and their laws would be the reasoned. But such a reversal at this level of attained knowledge would yield no additional knowledge. Knowledge before and after the reversal would be the same, so that the whole reversal would end up in a cognitively closed circle, and thus would be a mere *formality*. The type of reasoning given here yields knowledge about the ground I therefore characterize by the category *formal ground*.

The *second* type of knowledge is given in reasoning where the very ground is the reasoned by means of knowledge which is already independent from the knowledge about phenomena and their laws. The former type of reasoning about the ground is at the same time the presupposition for the latter type reasoning about the ground; the latter yields a knowledge about the ground which I characterize by the category *real ground*. By means of these two types of the category *ground* I can approach the movement of knowledge from Rutherford's thought construction of the atom to that of Bohr. While the former dealt with the inner structure of the atom *with respect* to the phenomenon of deflection of particles bombarding the atom, thus, his thought operations can be characterized by the category *formal ground*, the latter dealt in mind with the processes in the *very atom*, so his thought operations can be characterized by the category *real ground*. As I shall show now, the latter stands for knowledge which can be characterized by the category *working of the ground's mechanism*.

2.5 Bohr's Stationary States and the Stability of the Atom

Bohr in Part I of his article (1913b), which I analyze here, sets as his aim to deal with "the mechanism of the binding of electrons by a positive nucleus" (1913b, 2). He conceptually approaches this mechanism by means of Planck's view of radiation which he employs drawing on the "general acknowledgment of the inadequacy of classical electrodynamics in describing the behavior of system of atomic size" (1913b, 2). At the same time, Bohr clearly differentiates in his approach two distinct but still interrelated phases, *one*, in which he will provide "a basis for a theory of the constitution of atoms" (1913b, 2), and another one, based on the first one, in which he will give a "an account … for the law of the line spectra of hydrogen" (1913b, 2).

In Part I Bohr chooses as the object of his thought operations "a simple system consisting of a positively charged nucleus of very small dimensions and an electron describing closed orbits around it" (1913b, 3), and where this object is subjected to two additional suppositions: the mass of the electron is negligible small as compared to that of the nucleus, and the velocity v of the electron is much smaller than the velocity of light c. That choice enables him to bypass the issue of mechanical instability, and, when unified with the first supposition, it provides the conceptual framework for the treatment of the hydrogen atom. The latter supposition, characterized by means of the epistemological category *idealization* dealt with in Chapter 1, states that $\frac{v}{c} = 0$ holds. By the introduction of this idealization Bohr can abstract in mind from the impact of relativistic effects.

Bohr initially presupposes that no energy is radiated; thus he can apply classical mechanics to the case of the movement of the electron, on an elliptical orbit, around the nucleus located in the focus of this orbit. If e and m stand for the charge and mass of the orbiting electron, E for the charge of the nucleus, and $2a$ for the major axis of the orbit, then it holds (1913b, 3)

$$\omega = \frac{\sqrt{2}}{\pi} \cdot \frac{W^{\frac{3}{2}}}{eE\sqrt{m}}, 2a = \frac{eE}{W} \qquad (2.10)$$

Here ω stands for the orbital frequency of the electron, while W stands for the energy to remove it from the vicinity of the nucleus to infinity.

Then, Bohr gives up the supposition that energy is not radiated; this requires stabilizing the atom radioactively. To achieve this stabilization, he brings in Planck's idea of a discrete energy being radiated by a vibrator with the frequency v, and where the energy radiated in one emission is $\tau h v$, where τ is positive whole number and h Planck's universal constant. He applies this idea of Planck in such a way that he now supposes the electron, being initially at a great distance from the nucleus, was bound by the latter and settled in an orbit around it. To this binding Bohr applies Planck's idea in such a way that the energy being emitted in this process is given by the relation $E = \tau h \dfrac{\omega}{2}$, where ω stands for the orbital frequency of the electron on its final orbit in which it settles.[27] This relation, when combined with relations given in (2.10), yields –

$$W = \frac{2\pi^2 me^2 E^2}{\tau^2 h^2}, \omega = \frac{4\pi^2 me^2 E^2}{\tau^3 h^3}, 2a = \frac{\tau^2 h^2}{2\pi^2 meE} \qquad (2.11)$$

Bohr interprets these relations in such a way that for the respective values of τ one obtains the values for W, ω, and $2a$ which characterize the configurations of atom when no energy is radiated by the atom. These configurations he views as characterizing the states of the atom; so he assigns to them the term "stationary states." By bringing in the values for e, m, and h, and by supposing that $e = E$, he can for $\tau = 1$ he computes the linear dimension $2a$ of the atom of hydrogen, its optical frequency ω, and its ionization potential $\dfrac{W}{e}$; all this computed values correspond to those which computed on the basis of experiments. At the end of his treatment of the very atom, Bohr gives the following brief summary of his main assumptions (1913b, 7):

(1) The dynamical equilibrium of the systems in the stationary states can be discussed by help of the ordinary mechanics, while the passing of the systems between different stationary states cannot be treated on that basis.

(2) That the latter is followed by the emission of a homogeneous radiation, for which the relation between the frequency and the amount of energy emitted is the one given by Planck's theory.

---

27 Bohr uses in Part I two additional applications of Planck's view on quanta of energy; I do not deal with them in this chapter. On this see (Heilbron and Kuhn 1969).

What *epistemological conclusions* can be drawn from Bohr's treatment of the constitution of the atom? In my view, to Bohr treatment of the constitution of the hydrogen atom one can assign the category *real ground* understood as *working of the ground's mechanism*. This category involves as subordinated those categories which are given in (Machamer and Darden and Craver 2000) and (Darden 2008), once they are viewed as epistemological-*cum*-ontological categories. In (Machamer and Darden and Craver 2000) one can identify the following trio of categories falling under the category *working of the mechanism*: *start-up/set-up conditions, intermediate activities,* and *terminating activities.* The first category involves the following subordinated categories: *entities, relevant properties of entities given at the beginning of the activities, the mutual interactions of these entities, entities and activities as mutually correlative, the spatial distribution, orientation, and relations of the entities and activities, the temporal characteristics of the activities.* To the category *intermediate activities* are subordinated categories *intervening entities and activities,* understood as *producers of new states in entities and/or of new entities and activities.* The third category involves categories *endpoint of activities, parameters and state of the mechanism.*

This trio of categories with their subordinated categories, all falling under the category *real ground*, can be unified with the set of categories given by Darden (2008, 965) as follows:[28]

Table 2.5: *Categories falling involved in the category cluster* real ground

Main categories / Subordinated Categories	Components	Spatial arrangement of components	Temporal aspects of components	Contextual locations
	set-up/start-up & terminating conditions, entities, initial & intermediate & terminating activities, modules	localization structure orientation connectivity compartmentalization	order rate duration frequency	location within a hierarchy location within a series

---

28 The category *module* stands for reasoning about groups of components of the mechanism.

63

The set of categories given in this table, however, is not rich enough to express the central feature of Bohr's reasoning in Part I of (1913b), namely, that he arrives at the understanding of why a hydrogen atom endures. This endurance is based on unification in the mind of three features of the entities and activities involved in his thinking about the mechanism of the hydrogen atom: (a) the electron moves only in certain orbits; (b) these orbits are closed, that is, the termination conditions of electron's movement on the orbit are also its start-up conditions; and (c) there is no radiation due to the movement of the electron on this orbit. Thus, from the point of view of categories, Bohr stabilizes his thought construct by thinking its activities as *cyclical* activities.[29] Based on this feature, Bohr's thinking about the mechanism of the hydrogen atom is that about a *ground's mechanism which reproduces itself*. So as the ground is here thought by means of its mechanism's constituent and self-reproducing entities and activities, the ground's mechanism is the *reason* and the ground is the *reasoned*; the category *ground* is here is already the *real ground*. By providing the cluster of philosophical categories given above and pertaining to the mechanism of the ground, it becomes readily seen that *one can think the ground of phenomena and of phenomenal laws after, and at the same time, in a relatively independent way from thinking about them, and also prior to their "back" derivation.*

2.6 The Spectra Explained and Predicted

After Bohr dealt with the working of the mechanism of the hydrogen atom, he moved on to the derivation of the formula for the spectral lines of hydrogen. As shown above, Rydberg's formula (2.7) holds for the spectral lines of hydrogen. It can be transformed as follows. The relation $\lambda = c \cdot T$, where c is the speed of light and T the period of its movement, is transformed into $n = \dfrac{1}{c \cdot T}$, and by introducing the magnitude frequency v as $v = \dfrac{1}{T}$ one obtains $v = c \cdot n$. So, by multiplying the formula (2.7) by the magnitude c, one obtains

$$v = c \cdot N_0 \left(\frac{1}{n^2} - \frac{1}{m^2} \right) \qquad (2.12)$$

---

[29] For a philosophical treatment of thinking about cyclical processes in living systems see (Bechtel 2011).

For $c = 3 \cdot 10^{10}$ cms^{-1} and $N_0 = 109675.0$ cm^{-1} one obtains $c \cdot N_0 = 3.29 \ 10^{15}$ s^{-1}. Bohr compared this value, obtained by computation from the historically prior form of the formula for hydrogen, with the value he computed on the basis of the formula he derived from his theory about the hydrogen atom. He proceeded in the following way. He interprets the experimental process of discharge in vacuum tube when the spectrum is experimentally produced as a process of moving the orbital electron to a great distance from the nucleus. This latter process he views as corresponding to the creation of a stationary state when the electron is bound by the nucleus. For the energy W radiated in this binding holds the relation (2.11), and so as the object of his thought derivations is still hydrogen, he puts the absolute values of the charge of the orbiting electron and that of the nucleus as identical. He thus obtains from (2.11) the relation

$$W_\tau = \frac{2\pi^2 m e^4}{\tau^2 h^2} \tag{2.13}$$

Then, the energy produced when the atom passes from the state characterized by τ_1 to that characterized by τ_2 is as follows:

$$W_{\tau 2} - W_{\tau 1} = \frac{2\pi^2 m e^4}{h^2}(\frac{1}{\tau_2^2} - \frac{1}{\tau_1^2}) \tag{2.14}$$

This energy he puts equal to $h\nu$, where ν is the frequency of radiation, so he obtains

$$\nu = \frac{2\pi^2 m e^4}{h^3}(\frac{1}{\tau_2^2} - \frac{1}{\tau_1^2}) \tag{2.15}$$

From here Bohr's thought movement goes into three directions. One deals with the expression $\frac{2\pi^2 m e^4}{h^3}$ in (2.15), the second with the expression in the brackets of (2.15), and third one, on the basis of the whole formula (2.15), with Pickering series. The expression $\frac{2\pi^2 m e^4}{h^3}$ is used by Bohr for the recomputation of the constant given above as $c \cdot N_0$; for the values of e, m, and h it yields the value $3.10 \ 10^{15}$ s^{-1}. The expression in the brackets is used by Bohr in the *explanation* of already known spectral series and in the *prediction* of the existence of spectral lines as yet undetected. By putting $\tau_2 = 2$ and for a varying τ_1 he explains Balmer's series and for $\tau_2 = 3$ he *explains* Paschen's series. He *predicts*, by putting $\tau_2 = 1$, or 4, or 5, the existence of spectral series both in the extreme ultra-violet and the extreme ultra-red, which "are not observed, but the existence of which may be expected" (1913b, 8). Already in 1914 the series corresponding to $\tau_2 = 1$ was detected; few years later the series for τ_2 equal to 4 and 5 were also experimentally detected.[30]

The whole formula (2.15) is applied by Bohr in the reinterpretation of the Pickering series assigned to hydrogen. So as the positive whole number values of variables given in the bracket of (2.15) are interpreted by Bohr as ordering numbers of the orbits of the electron in the hydrogen, the value ½ given in the formula for Pickering series could not be assigned any physical meaning in Bohr's theory of hydrogen. Bohr therefore, based on the formula (2.15), reinterprets the formula for Pickering series by means of the following two transformations:

$$v = \frac{2\pi^2 me^4}{h^3} \left(\frac{1}{(\frac{\tau_2}{2})^2} - \frac{1}{(\frac{\tau_1}{2})^2} \right)$$

This formula can be rewritten as

$$v = \frac{2\pi^2 me^2 (2e)^2}{h^3} \left(\frac{1}{\tau_2^2} - \frac{1}{\tau_1^2} \right) \qquad (2.16)$$

where "$(2e)^2$" is viewed as corresponding to "E^2" in (2.11), that is, to the contribution of a nucleus' charge equal to the charge of two electrons. An atom with such a nucleus is classified by Bohr as that of helium, and so as in (2.16) "e^2" stands for the presence of just one orbital electron, the atom of helium is here understood as an ionized one. Based on his theory, Bohr interprets the Pickering series as pertaining to helium and not to hydrogen. Drawing on this interpretation he can also explain why this series was initially viewed as that of hydrogen (1913b, 10):

> If we put $\tau_2 = 4$, we get the series observed by Pickering in the spectrum of ζ Puppis. Every second of the lines in this series is identical with a line in the Balmer series of the hydrogen spectrum; the presence of hydrogen in the star in question may therefore account for the fact that that these lines are of a great intensity than the rest of the lines in the series.

---

[30] On this see (Lyman 1914), (Brackett 1922) and (Pfund 1924). I do not deal here with the confirmation of Bohr's theory on the basis of the experiments of Franck and Hertz because they pertained to vapors of mercury, that is, to a many-electron atom and not to the hydrogen atom as a single-electron atom with which I dealt here. On this see (Franck and Hertz 1914; 1919). Bohr's interpretation of the results of their experiments is given in (Bohr 1915), for an interpretation see also (Van de Bijl 1917a; 1917b).

Evans in (1913), by explicitly drawing on Bohr's interpretation of the Pickering series, performed experiments with tubes containing helium from which hydrogen was eliminated; their results confirmed this interpretation.

Finally, still in Part I, Bohr leaves the framework of reflections on hydrogen atom and its spectra, and gives an interpretation of Ritz combination principle as holding for all substances producing line spectra (1913b, 11):

> The circumstance that the frequency can be written as a difference between two functions of entire numbers suggests an origin of the lines in the spectra in question similar to the one we have suggested for hydrogen, i.e. that the lines correspond to a radiation emitted during the passing of the system between two stationary states. This may account for the different sets of series in the line spectra emitted from substances in question.

What *epistemological lessons* can be drawn here? Once Bohr succeeded in the above reconstructed three directions of though movement, his *knowledge has grown*. What I mean by this can be understood when I put side by side the initial formulation (2.12) of the spectral formula for hydrogen and the spectral formula (2.15) derived by Bohr.

$$\upsilon = c \cdot N_0 (\frac{1}{n^2} - \frac{1}{m^2}), \; \upsilon = \frac{2\pi^2 m e^4}{h^3} (\frac{1}{\tau_2^2} - \frac{1}{\tau_1^2}) \qquad (2.17)$$

By their comparison it becomes readily seen that the expression "$c \cdot N_0$" is restated as $\frac{2\pi^2 m e^4}{h^3}$, and so the formula for the phenomenon of spectral emission is completely tied to its explanans, the latter being Bohr's unification of Rutherford's atom with Planck's views on the quanta of radiation. The symbols "e," "h," and "m," together with their respective meanings appear in the explananda and predicanda because there are shifted here in the process of explanation and prediction from Bohr's explanans – the understanding of the ground's mechanism producing the respective phenomena.[31] By this comparison one finds out also that the interpretation of the symbols "n" and "m" is different from that of "τ_2" and "τ_1." While the former pair stands just for an ordered pair of numbers, the latter stands for the sequence of orbits of the electron.

---

31 In order not to overburden the text I use here only the category *explanans* as the basis for the thought derivation of both the explananda and predicanda; thus refraining from the use of the category *predicans*.

The growth of knowledge becomes readily seen also when one reconstructs the respective scientific laws into which the formulas (2.17) are embedded. I reconstruct them as follows:

$$\forall x[H(x) \& I(x) \rightarrow \upsilon(x) = c \cdot N_0 (\frac{1}{n^2} - \frac{1}{m^2})],$$

$$\forall x \forall y[El(x) \& N(y,x) \& H*(z,x,y) \& D(z) \& \frac{v(y)}{c} = 0 \rightarrow \upsilon(x)$$

$$= \frac{2\pi^2 m(x)^{(1)} E(y)^2 e(x)^2}{h^3} (\frac{1}{\tau_2(x)^2} - \frac{1}{\tau_1(x)^2})]$$

In the antecedent of the second law "$El(x)$" stands for "x is an electron," "$N(y,x)$" for "y is the nucleus orbited by x", "$H(z,x,y)$" for "z is hydrogen composed of x and y", "$D(z)$" for "z is disturbed," and "$\frac{v(y)}{c}$" for "the ratio of the speed of y and speed of light." In the consequent "$m^{(1)}(x)$" stands for "mass of x subjected to one idealization,"

By comparing these two laws, one discovers that what has changed is the very meaning of the term "hydrogen," that is, the universe of discourse to which the law can be applied at all. While in the former law, symbolized as "H," it is understood as, say, a substance with a certain atomic weight, reacting in certain proportion with other substances and with a certain ionization potential, in the latter, symbolized as "$H*$," it is understood as being set up by an electron orbiting the nucleus. What has changed also is the understanding when the hydrogen is at all radiating: in the former law it has to be incandesced, in the latter law it has to be disturbed from the outside by an input of energy, so that the electron is shifted from one orbit to another. Finally, Bohr's equation given in the law holds only when the idealization holds that the orbiting speed of the electron inside the hydrogen must be much smaller than that of light.

Another feature of the growth of knowledge can be understood once one philosophically reflects on Bohr's reinterpretation of Pickering series. Initially, the Balmer and the Pickering series were assigned as *two phenomenal laws to the same system* – the hydrogen substance. Based on Bohr's understanding of the system's mechanism generating spectral lines, these laws were reinterpreted in such a way that they were assigned to two different systems each of them with its own specific mechanism: the hydrogen and the helium.

Based on these analysis, I come to the conclusion that while it is possible to assign both to the knowledge which is the point of departure of the thought movement *to* the ground's mechanism as well as to the knowledge which is obtained by explanation and prediction *from* the knowledge about the ground's mechanism the same category pair *phenomena* and *law of phenomena*, this pair of philosophical categories does not grasp the above reconstructed cases of growth of knowledge. The phenomena and laws of phenomena as the epistemic points of departure for the movement to ground's mechanism and as the latter's epistemic consequences differ mutually in some fundamental aspects. This difference, as can be seen from Figure 2.1, was taken into account neither in the article (Bogen and Woodward 1988), nor in the ensuing discussions on categories *data* and *phenomena*. As a consequence of the absence of this account, the whole cycle *data* → *phenomena* → ... → *phenomena*, reconstructed for the first time in (Bogen – Woodward 1988), cannot serve the epistemological purpose of grasping the extension of knowledge. A remedy to this I shall give in the concluding part of this chapter. This failure could then be turned into an argument arguing in favor of a modified logico-empiricist view, namely, that *even if the phenomena are not observable, still a detour through theory and the knowledge of mechanism given in it does not make any sense*; one starts and ends with the same knowledge expressed by the category phenomena. A remedy to this deficit will be given below.

Another epistemological conclusion which can be draw here pertains to the issue of *observability* and *perceptibility* in general. The frequencies v both explained a predicted by Bohr do not refer to something accessible to sensory experience, as can be seen from the following table stating the value of the frequency of the first line in each series:

Table 2.6: *The nonobservable character of frequencies of spectral lines explained/predicted by Bohr*

Characteristics of the series / Series	τ_2	τ_1	v 10^{12} sec^{-1}
Predicted	1	2	2325
Explained	2	3	430
Explained	3	4	150
Predicted	4	5	69
Predicted	5	6	37

Thus, neither the phenomena and laws of phenomena as the epistemic points of departure for the movement to ground's mechanism nor the phenomena and laws of phenomena as derived from the understanding of the ground's mechanism *refer to state of affairs accessible to human sensory perception*.

Worth noting here, from the point of view of the issue of observability, is also that when Bohr compares the computed value of the expression $\frac{2\pi^2 me^4}{h^3}$ with the computed value of $c \cdot N_0$ equal to $3.290 \cdot 10^{15}$ cycles per second, he assigns to the former value the predicate "theoretical" and to the latter the predicate "observed," declaring that the "agreement between the theoretical and observed values is inside the uncertainty due to the experimental errors in the constants entering into the expression for the theoretical value" (1913b, 9). In my view this assignment of epistemological predicates does not hold here. *First*, no human being has ever *observed* a frequency with $3.290 \cdot 10^{15}$ cycles per second. *Second*, the very magnitude frequency is not observable in the sense of being accessible to human perceptual systems. It can only be derived in thought based on the knowledge of the ratio $\frac{c}{\lambda}$, and, of course, no one has ever perceived a velocity in the range of 10^{10} centimeters per second or the very magnitude velocity. *Third*, as shown above in the case of Ångström's computations, the magnitude wave length λ is not perceived; it has to be computed on the basis of knowledge of the grating space, of the angle of diffraction while taking into account also the correction factors like the variation of temperature and the orientation of the grating. *Fourth*, the value of the constant N_0 entering into the computation yielding the result $3.290 \cdot 10^{15}$ cycles per second, cannot be perceived. As shown above, it was computed by Rydberg via the relation $\frac{4}{\lambda_0}$, that is, on the basis of Balmer's basic wave length λ_0, which – in turn – was obtained from Ångström's computations of the wave lengths of the hydrogen lines; the latter however are due to their size not observable but only computable.

Once the magnitudes like λ, c and υ, as well as their actual values assigned to the spectral lines are not perceptible, but are in fact derived from various formulas by means of computations, then to all of them has to be assigned, form the point of view of philosophy of science, the epistemic status of *thought things*, *things in mind*, on which one can perform thought operation and obtain in such a way other magnitudes and their actual values. A backward look at my reconstruction of the development of physics from Ångström to Bohr shows that what was here involved in addition to thought operations were also *practical operations* with things of nature. The relation between these two types of operations I shall now examine from the point of view of philosophy of science.

2.7 General Epistemological Lessons

Based on the above given reconstructions of the growth of knowledge, I propose the following, more fine grained differentiation between philosophical categories. Knowledge which was till now labeled by the category *phenomena* is now labeled either by the category *appearance* or by the category *manifestation*, and knowledge which was till now labeled by the category *law of phenomena* is now labeled either by the category *law of appearance* or by the category *law of manifestation*. The philosophical category *appearance* is assigned to that level of knowledge and thinking where the phenomena are thought as yet independently and prior to their unification in scientific laws, where the latter have the status of *laws of appearance*; it holds for them the characterizations of the category *law of phenomena* given above. The category *law of appearance* characterizes knowledge and thinking before the knowledge about the ground's mechanism is derived from them. Once the very mechanism of the ground is understood, the laws of phenomena and the phenomena derived (explained and/or predicted) on its basis have already the status of *laws of manifestation* and of *manifestations*. The category *law of manifestation* can be characterized as follow: *it unifies in mind a number of different phenomena pertaining to entities of the same kind, and where this unification stands for the derivation based on the understanding how the phenomena are produced by the working of the mechanism of the ground of these phenomena.*

How can the difference between categories *appearance* and *law of appearance*, on the one hand, and *manifestation* and *law of manifestation*, on the other hand, be epistemologically explicated? As shown above, the laws of spectra and thus also the frequencies computed on their basis, once derived from Bohr's understanding of the working of the mechanism of the atom, contain already the symbols e, m, h, τ_1, and τ_2. These symbols, as well as their respective meanings, are part of the explananda/predicanda and have their origin in their common explanans – Bohr's understanding of the working of the mechanism, which is based on the meaning of such terms as "electron moving on in orbit above the nucleus," "quantum of radiation emitted," "stationary orbits," etc. This can be stated in more general, epistemological terms as follows. Concepts introduced at the level of knowledge and thinking characterized by the category *working of the mechanism of the ground* – as the explanans – are shifted in the process of explanation/prediction to the explanandum/predicandum. So, while at the level of knowledge and thinking characterized by the categories *appearances* and *laws of appearances*, concepts standing for the knowledge about the working of the mechanism are as yet not given, they are already given at the level knowledge characterized by the categories *laws of manifestations* and *manifestations*. This is the difference between knowledge characterized by the categories *appearance and laws of appearance* and knowledge characterized by the categories *law of manifestation* and *manifestation*.

From this point of view I can provide yet another explication of the difference between knowledge characterized by the category *formal ground* and knowledge characterized by the category *real ground*. In the former, as shown above, the reversal turning the knowledge of the ground from the reasoned into the reason, and of the phenomena and of their laws from the reason into the reasoned, yields no growth of knowledge, that is, the phenomena and their laws before *and* after the reversal have still the same cognitive status – they can be characterized by the same philosophical categories: *appearances* and *laws of appearance*.

Contrary to this, knowledge characterized by the category real ground provides the basis for the "back" derivation of the appearances and laws of appearance as manifestations and laws of manifestation. Thus, the turn of the knowledge of the ground from the reasoned to the reason, and of the phenomena and of their laws from the reason into the reasoned lost its formal character: these turns yield already an extension of knowledge. This type of turn in mind can be characterized also as a *non-formal type of the retreat into the ground and coming out of it*.

Based on this differentiation of categories, I propose – as an alternative to S. Schindler's Figure 2.1 – the following representation of the sequence of categories reconstructed till now.

Figure 2.2: Sequence of categories reconstructing the growth of scientific knowledge

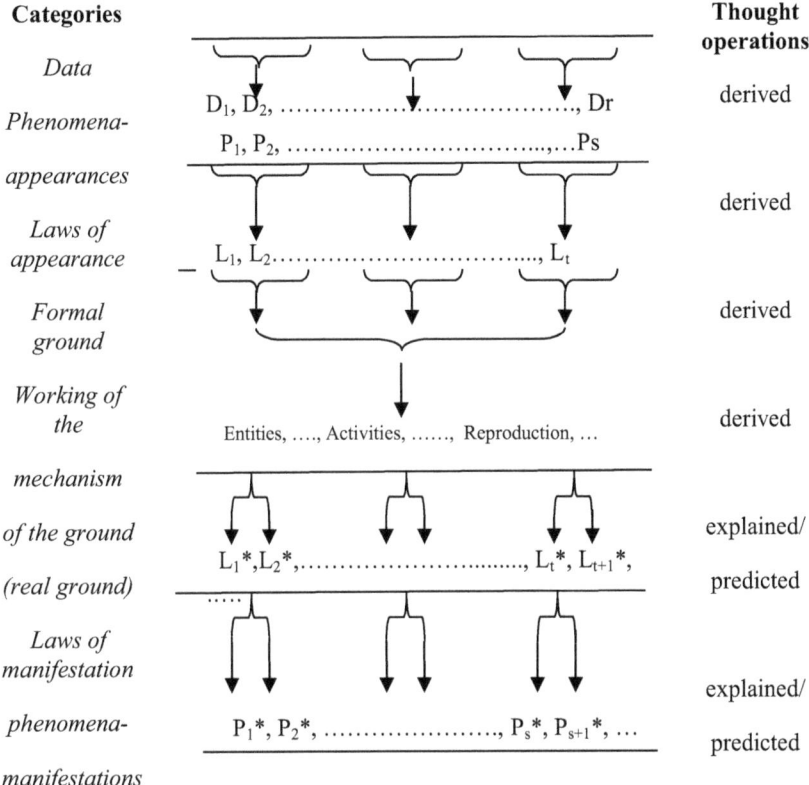

Here the scientific laws $L_1^*, L_2^*, ..., L_t^*$ stand for a reinterpretation by means of explanation of the scientific laws $L_1, L_2, ..., L_t$, while the laws $L_{t+1}^*, ...,$ stand for laws of manifestation which do not have their counterpart in the laws of appearance; they are *predicanda* laws and not explananda laws. The same holds for manifestations expressed as $P_1^*, P_2^*, ..., P_s^*$; they are the reinterpretations, based on explanation from the laws of manifestation, of the appearances $P_1, P_2, ..., P_s$, while the manifestations $P_{s+1}^*, ...,$ are *predicanda phenomena* derived from the laws of manifestations. As an alternative to Bogen's and Woodward's approach to categories *data* and *phenomena*, summarized in Table 2.1, I propose Table 2.7; its reading should be from the cells of the first line to the cells in the first column left.

Table 2.7: Relations between categories data, phenomena, etc.

Thought-movements from / to	Data	Phenomena as appearances	Laws of appearance	Mechanism of the ground	Laws of manifestation	Phenomena as manifestations
Data	x	x	x	x	x	x
Phenomena as appearances	basis for inference of	x	explanatory/ predictive basis for inference of	x	x	reinterpret
Laws of appearance	x	basis for inference of	x	x	reinterpret	x
Mechanism of the ground	x	x	basis for inference of (formal ground)	real ground	x	x
Laws of manifestation	x	x	reinterpreted as	explanatory/pre-dictive basis for inference of	x	x
Phenomena as manifestations	x	reinterpreted as	x	x	explanatory/ predictive basis for inference of	x

Based on above given reconstruction of the development of physics I propose the following. The whole sequence represented in Figure 2.2 stands for an interchaining of thought operations which are completely separated from sensory perception; the level of knowledge characterized by the category data is in my approach related not to perception but to practical operations in the world, e.g., as shown in Ångström's experiments and computations of the wave lengths of spectral lines, to experiments in a laboratory.

How can these operations be epistemologically characterized in more detailed way? I characterize them by means of the categories given in Table 2.5, Figure 2.2, and Table 2.7.

My employment of categories like *thought object* or *thought thing* and *thought operations*, together with the explication of categories given in Figure 2.2 and Tables 2.4, 2.5 and 2.7, can at the same time be viewed as an epistemological explication of the claim: "Intelligibility, at least in molecular biology and neurobiology, is provided by descriptions of mechanisms, that is, through the elaboration of constituent entities and activities that ... provide an understanding of how some phenomena are produced" (Machamer and Darden and Cramer 2000, 21). Stated in even more general epistemological terms: thought operations on things of mind stand for the production of net of thought determinations by means of which we make the natural world intelligible to us. Categories like *data, phenomena, mechanism*, etc., can be viewed as philosophical reconstructions of the central "knots" of this net. These categories have therefore an *epistemological* and at the same time *metaphysical* status, and because they reconstruct the operations performed by thinking on the things in mind, these categories in their unity stand for a *logic* of scientific thinking.

My approach to thinking as presented here is in an explicit opposition to the view presented by J. W. McAllister. He, at least in my view, correctly claims that empirical science "aims to ascertain the structure of the world" (2011, 73), and that it does so "by analyzing empirical data, consisting of the outcomes of observations and measurements" (2011, 73). But based on this, he draws the wrong conclusion that the outcomes of observations and measurements "constitute our sole epistemic access to reality" (2011, 73). In my approach, not only the thought appropriation of the outcomes of experiments as data, but *all* the thought transformations and their results symbolized in Figure 2.2 stand for an ascertaining of the structure of the world.

From this point of view one can identify the following contradiction in the works of P. Machamer. On the one hand, he was able to provide a path breaking philosophical reconstruction, by means of categories, of thinking about mechanism, which by its very activity and results is not related to perception. Thus, for example, he claims that "knowledge ... is not passive: knowledge representations are not static traces deposited by incoming signals, but active representations that must include activities on the part of the knower ... That is, acting is a major part of knowing" (Machamer 2004, 33). But once he moves to the *metalevel* – with respect to the categories he assigns to thinking about mechanism – he gives epistemical priority to perception by drawing on the article (Scholl and Tremulet 2000), which he interprets as follows (Machamer 2004, 33):

> [T]here is strong evidence to show that what are often taken to be interpretations of causality and intentionality resulting from perceptions are themselves perceptual phenomena that are brought about by perceptual modul innate in the human visual system. Stressing [the ...] perceptual nature [of schemata and categories of the productive activity] is supposed to distinguish them from inferences or from higher level cognition. The fundamental claim stresses how they are in fact the way human epistemologically [read "epistemically"] access and make sense of the world.

In my approach, all types and levels of thinking, characterized by the category cycle *data* → *phenomena-appearances* → ... → *phenomena -manifestations* represented in Figure 2.3, have nothing to do with perceptions or perceptual modules, but still they stand for an *epistemic appropriation of the world*. So, the question *where does the awareness about world of things come from* is answered by me here, that is, in the framework of epistemology as follows: *one form and level of awareness comes from and originates in other forms and levels of awareness*.[32]

Thus far, I have dealt only with one aspect or "dimension" of thinking, namely, the reproduction of a thing in the world outside the mind as a thought thing (thing in the mind). Human thinking is, however, a process which in itself relates to the thought world of things in mind and at the same time to the (possible) thinking of other humans about the world, that is, takes its course in ways which are *objectively* and in the same time *intersubjectively* valid. This thinking has the specific feature that it can think the supraindividual space where the mutual mediation of thought processes of human beings can take place.[33] The objective content of thinking thus serves not only the purpose of the discovery and more precise determination of the characteristics of the natural world but is also used in the exchange of thoughts via communication. Therefore one individual can interlock his/her own thought operations to the result of thought operations of

---

32 The question about the ontogeny of awareness in human infants and the question about the phylogeny of awareness are the subject matter of investigation not by epistemology but primarily by developmental psychology, evolutionary anthropology and paleoanthropology.
33 On this see (Jantzen 2003).

other humans; e.g., Bohr appropriated in mind the results of analysis of other spectroscopists as well as the thought constructs of atoms created by J. J. Thomson and E. Rutherford. Instead of speaking about the *intentionality* of thinking, I speak – following (Tomasello *et al.* 2005) – about the *I-* or *shared-intentionality* of human thinking; due the latter characteristics of thinking, the conceptual innovations performed by an individual can be passed on to other individuals, that is, can be socially inherited. Thus, when I employ in this chapter the category *mind* I mean by it *cooperated mind*, by the categories *thinking* and *cognition* I mean *intersubjective activities*, and by the category *knowledge* I mean *intersubjectively* given knowledge. These issues will be treated in Part III of this book.

Chapter 3: Observability and Theory-Ladenness of Observation: Myths and Facts

This chapter analyzes issues which can be regarded as central for both philosophy and methodology of science, as well as for epistemology, namely, observability and theory-ladenness of observation.

I will start with an overview of the approach taken by the logical positivists of the early 1930's to the issue of observability and indicate how this approach changed in the framework of logical empiricism from the mid-1930's until its final demise in the late 1960's and early 1970's. Then, I shall deal briefly with the issues of observability and theory-ladenness of observation as understood by P. K. Feyerabend in the 1960's.

Instead of dealing with the more recent philosophical analyses,[34] my intra-philosophical approach to these two issues will stop here (i.e., with Feyerabend's analysis), because I want to take another approach to them, by dealing with a specific period in the history of science. Here I mean the formulation of Bohr's theory of the hydrogen atom dealt with in Chapter 2, which I employ here as a case study upon which, then, will be based my proposal of an epistemological generalization and, finally, a resolution of problems surrounding the issues of observability, theory-ladenness of observability, as well the place of practical operations in science.

3.1 Logical Positivism/Empiricism and the Post-positivistic Backlash: The Myths

3.1.1 From Logical Positivism to Logical Empiricism

Logical positivism from the late 1920's to the mid 1930's can be viewed as an attempt to blend Mach's sensualism (as a variety of phenomenalism) with modern symbolic logic originating in the works of Frege and Russell, while at the same time to some extent bringing in the views presented in Wittgenstein's *Tractatus*.

In a self-reflection it was declared that logical positivism stands for "the convergence of two significant traditions: the positivistic-empirical and the logical"

---

34 On this see, for example, (Brewer and Lambert 2001), (Mehmet 2008), (Rochefort-Maranda 2008), (Schindler 2013).

(Blumberg – Feigl 1931, 281) upon whose bases is created a unified epistemology "in which neither logical nor epistemological factors are neglected" (Blumberg – Feigl 1931, 282). In this epistemology (Blumberg – Feigl 1931, 286):

> The attempt is made to show ... that all concepts of empirical science can be constituted or constructed by purely logical operations upon a single primitive relation and the primitive elements between which it holds ... the terms between which the relation subsists are total momentary experiences, cross sections of consciousness. This means in short that any proposition of empirical science can be completely translated into a complicated series of propositions which refer only to the relational structure of the given.

With respect to the understanding of this "given" it seems obvious that logical positivism, from the point of view of *epistemology*, was a direct heir of Mach. This heritage was characterized as follows (Toulmin 1969, 33–34):

> Above all, Ernst Mach was to be godfather of logical positivism ... All claims to knowledge of the world around us, Mach argued, derived their justification from the evidence of our senses, and this "evidence" must ultimately be interpreted in terms of the direct content of our individual sense fields. Accordingly, the theory of knowledge ... was reduced for him to ... the analysis of sensations. Mach's epistemological ... was a "sensualist" one.

Mach's epistemology was then "logicized" once it was declared that the counterparts of elementary sense data (atomic facts) were the so-called "Protokollsätze" to which all other sentences of the total language of empirical science should be reducible.[35] According to logical positivism, "every descriptive term of science could be defined by perception terms, and hence, that every sentence of the language of science could be translated into a sentence about perceptions" (Carnap 1936/37, 464).

Starting from the mid-1930's, as can be seen in Carnap's *Testability and Meaning*, a process of gradual relaxation of the radical sensualism was initiated. Instead of a complete testing-verification of theories (scientific laws) by a *complete* reduction of their statements to the observable, Carnap speaks about *confirmability* and acknowledges that theories (scientific laws) cannot be *completely eliminated in favor of the level of directly observables* (1936/37, 420–429, 463). These views were then further developed in his works (1939; 1956; 1958; 1959/2000), as well as in Hempel's (1951; 1952; 1958) and in Feigl's (1956).

In the *Foundations of Logic and Mathematics* Carnap presented several of the central characteristics of the so-called "standard conception" of scientific theories. The latter were viewed as being composed of a logico-mathematical calculus

---

35 On the "Protokollsätze" see (Carnap 1932a; 1932b).

unified with axioms and postulates of the respective science (e.g., physics) which are viewed as purely syntactic entities, that is, as chains of uninterpreted symbols. The empirical meanings of these symbols are obtained once the theorems and particular statements derived from the axioms and postulates are semantically interpreted by assigning to them observationally given designates.

Worth noting here is that in Carnap's approach, a semantic, that is a designation-interpretation is given only to those symbols which designate, he claims, observables. Symbols given in theories like, e.g. "ψ" given in quantum mechanics and "$\bar{E}$" given in electrodynamics, are denied the status of designators once they lack observable designates. And when they are understood as, for example, "wave function" or "vector of electric field," this understanding does not stand for an assignment of designates to them, but only for a shift "from a symbol in a symbolic calculus to a corresponding word expression in a calculus of words" (1939, 68). The overall approach to scientific theories was expressed by Carnap as follows (1939, 65):[36]

> The calculus is first constructed floating in the air, so to speak; the construction begins at the top and then adds lower and lower levels. Finally, by the semantical rules, the lowest level is anchored at the solid ground of observable facts. The laws, whether general or special, are not directly interpreted, but only the singular sentences. For the more abstract terms, the rules determine only an *indirect interpretation*, which is ... incomplete in a certain sense.

In Carnap's ensuing works (1956; 1958, 1959/2000) these views were further developed by introducing the distinction between the theoretical language (L_T) and observational language (L_O) of science and where the so-called "bridge principles" (correspondence principles) mediate between these languages and thus enabling to derive observable consequences of a scientific theory. So, while the shift from logical positivism to logical empiricism was based on a departure from a radical sensualist position, still Carnap and his followers held to a empiricist position which was expressed by Hempel as follows (1958, 37, 41):

> Scientific research in its various branches seeks not merely to record particular occurrences in the world of our experience: it tries to discover regularities in the flux of events and thus to establish general laws which may be used for prediction, postdiction, and explanation ... Scientific systematization is ultimately aimed at establishing explanatory and predictive order among the bewilderingly complex "data" of our experience, the phenomena that can be directly "observed" by us.

---

36 For such a "spatial" view of scientific theories see also (Hempel 1952, 36) and (Feigl 1970a, 5–6).

The gradual departure from the views of the late 1920's and early 1930's can be retroactively interpreted as a process which resulted in an erosion of the empiricist core of the tenet of both logical positivism and logical empiricism. This erosion is readily seen when one compares the views of Hempel in (1950; 1951; 1952; 1970; 1974) with those of Feigl in (1956; 1970a; 1970b; 1974).

As early as 1950 Hempel declared that (p. 172)

> most scientific hypotheses do not ... assert any observation sentences at all ... that therefore their meaning cannot be adequately stated in terms of observation sentences alone ... In many cases, such a complex confirmatory relationship is based on subsidiary hypotheses which establish a connection between the hypothesis under test and the evidential basis ... Hence, any attempt to characterize the import of a scientific hypothesis exclusively in terms of potential observational evidence is bound to prove inadequate ... the meaning of a statement in the language of science is reflected in the totality of its logical relationships (those of entailment as well as those of confirmation) to all other sentences of that language. The logical relationships of the statement to the evidence sentences form only part of that complex relational network which represents its meaning.

And he declared also (1951, 68–69):

> It is not correct to speak ... of the "experiential meaning" of a term or a sentence in isolation. In the language of science ... a single statement has no experiential implications. A single sentence in a scientific theory, for example, does not, as a rule, entail any observation sentences, consequences asserting the occurrence of certain observable phenomena can be derived from it only by conjoining it with a set of other, subsidiary hypothesis.

Feigl also subscribed to such a holistic-cum-network view of scientific theories; "the meaning of theoretical constructs is best explained in terms of their locus in the nomological network" (1956, 17).

However, from this shared view they drew different and, in fact, opposing methodological and epistemological conclusions. In defending against both the sensualism of orthodox (logical) positivism and the post-positivist claim about the theory-ladenness of observation, Feigl differentiates between empirical lawful relationships among sense impressions having the status of common sense assumptions and the sense impressions themselves, so that the only the former provide the corroboratory basis for high level theories, and where these laws are independent from these theories, so that when they are tested, the empirical laws are not under scrutiny (1970b, 7; 1974, 8–9)

As an example of an empirical law which is not revised with respect to a higher level theory, Feigl mentions "the empirically discovered formula of Balmer (1885) for the order of lines in the hydrogen spectrum ... and thirty years later [it] became an important piece of empirical evidence for Bohr's quantum theory of the atom" (1974, 10). However, at the same time, he stated in respect to that common sense that "in an epistemological reconstruction of the common sense view of the world ... something akin to a sense data analysis is indispensable ... sense data, i.e., artificial reconstructions of the items of direct experience can serve as the substantiating evidence for our everyday knowledge of matter of facts" (1974, 7).

Hempel, unlike Feigl, in (1970; 1973) *completely gave up empiricism*. He declared (1970, 142–143):

> Theories are normally constructed only when prior research in a given field has yielded a body of knowledge that includes empirical generalizations or putative laws concerning the phenomena under study. A theory then aims at providing a deeper understanding by construing those phenomena as manifestations of certain underlying processes governed by laws which account for the uniformities previously studied, and which, as a rule, yield corrections and refinements of the putative laws by means of which these uniformities had been previously characterized.

So, according to him, one can distinguish two kinds of statements with their respective vocabularies: the vocabulary related the empirical generalizations/laws, the so-called "*pre-theoretical* or *antecedent* vocabulary relative to the theory in question" (1970, 143) and the vocabulary of the theory, that is, the *theoretical vocabulary*. The only difference between these two vocabularies is that the former "will ... be available and understood before the introduction of the theory, and its use will be governed by principles which, at least, initially, are independent of the theory" (1970, 143). At the same time he renounced the central claim of empiricism (1970, 143):

> The antecedently examined phenomena for which a theory is to account have often been conceived as being described, or at least describable, by means of an observational vocabulary, i.e., a set of terms standing for particular individuals or for general attributes which, under suitable conditions, are accessible to "direct observation" by human observers. But this conception has been found inadequate on several important counts.

The justification of this renouncement is follows (Hempel 1970, 143):

> the antecedent vocabulary need not and indeed should not, generally be conceived as observational ... for the antecedent vocabulary of a given theory will often contain terms which were originally introduced in the context of an earlier theory, and which are not observational in a narrow intuitive sense.

As an example he mentions Balmer's formula for the wave length of spectral lines of hydrogen, declaring that (1970, 145)

> terms such as 'hydrogen gas', 'wave length' ... are not observational terms ... the terms are antecedently understood ... for when Bohr proposed his theory of the hydrogen atom, principles of their use, including principles for the measurement of optical wave-

lengths, were already available, they were based on antecedent theories, including wave optics.

3.1.2 Post-positivism and the Theory Ladenness of Observation

Feyerabend, like Hanson and Kuhn, deviated in some crucial aspects even from the reformed views stated by logical empiricist in the period from the mid-1930's to the late 1950's.

Accordingly, the following claims of Feyerabend are worth quoting at length.

(1) (1958, 147)

> Now any philosopher who holds that scientific theories and other general assumptions are nothing but convenient means for the systematization of the data of our experience is thereby committed to the view ... that *interpretations ... do not depend upon the status of our theoretical knowledge* ... No independent interpretation is given for the theoretical terms ... This implies that the interpretation of a theory depends upon the interpretation of the observational-language used, but not the other way round.

(2) (1962, 29)

> The influence, upon our thinking, of a comprehensive scientific theory ... goes much deeper than admitted by those who would regard it as convenient scheme for the ordering of facts only ... scientific theories are ways of looking at the world; and their adoption affects our general beliefs and expectations, and thereby also our experiences and our conception of reality ... what is regarded as 'nature' at a particular time is *our own product in the sense that all the features ascribed to it have first been invented by us and then used for bringing order to our surrounding.*

(3) (1965, 180, 213)

> The meaning of every term we use depends upon the theoretical context in which it occurs. Words do not "mean" something in isolation; they obtain their meanings by being part of a theoretical system ... According to the point of view I am advocating, the meaning of observation sentences is determined by the theories with which they are connected. Theories are meaningful independent of observations; observational statements are not meaningful unless they have been connected with theories ... It is therefore the *observation sentence* that is in need of interpretation and *not* the theory.

From these statements Feyerabend draws as a conclusion the following radical anti-empiricist claim (1970, 92):

> It must be possible to imagine a natural science without sensory elements ... experience arises together with theoretical assumptions, not before them ... sensations can be eliminated from the process of understanding (though they may of course continue to accompany it, just as a headache accompanies deep thought).

If we accept this claim, then all scientific knowledge acquires the status of theory. But then the problem we face is *how a theory can be connected, if it can be connected at all, with nature*, with respect to which it should be tested. Can nature at all be accessible to scientists even if with an epistemic (theoretic) background, but still in a *non-epistemic (non-theoretic)* way? How can we escape the confines of a theory and of the linguistic framework in which it is stated? From this point of view, the sensualist strategy of logical positivism/empiricism, even if it is a failure, makes perfect sense: "pure" perception and/or "pure" (theoretically not contaminated) observations should enable one to escape the theoretical framework.

In subchapter 3.4 I will try to provide a different "escape" route. First, let me try to refute the myth that a theory cannot be tested by its consequences because they are loaded by the theory itself.

3.2 Can a Theory-Loaded Theory Be Tested? A Case Study

The standard argument for the impossibility of testing a theory via its consequences can be stated, seemingly, as follows:[37]

1. If our observation reports are theory laden, then our empirical tests are circular.
2. Our observation reports are theory laden.
3. Therefore, our empirical tests are circular.

This argument is restated by M. Hesse into the form of the following question: "If the use of all observation predicates carries theoretical implications, how can they be used in descriptions which are claimed to be evidence for the same theories?" (1974, 33). While in the argument one speaks generally about "theory-ladenness," Hesse speaks more precisely about "the same theory." As a result, I restate the argument as follows:

1. If our observation reports are loaded by a theory, then our empirical tests of this theory by means of these reports is circular.
2. Our observation reports are loaded by a theory.
3. Therefore, our empirical tests of this theory by means of these reports are circular.

Let me know try to show, by dealing briefly again with the relation of Balmer's pre-quantum spectral formula to Bohr's theory of hydrogen, that this argument and in fact the whole theory-ladenness-of-observation thesis is based on a superficial understanding of the process of theory-testing.

---

37 Here I draw on (Rocheford-Miranda 2008).

3.2.1 Balmer's Formula and Bohr's Hydrogen Atom

A.-J. Ångström published in (1868) the results of measurement of the wave lengths of lines in the solar spectrum. His aim was to produce a detailed atlas of the wavelengths of spectral lines given in the solar light by employing optical gratings.

The basic equation, based on the wave theory of light, used by Ångström for the computation of the wave length λ of the particular lines was $\lambda = e \cdot \sin\varphi$ for the spectrum of the first order, for its N-th order holds $\lambda = N \cdot e \cdot \sin\varphi$. Here e stands for the width of the grating space whose value was determined already in the production of the grating by means of a dividing engine. The magnitude φ stands for angle of diffraction of the light waves measured in the course of experiments with solar light and its actual values were found out by the employment of a theodolite. Then, Ångström modified that equation by taking into account the deviations of temperature t from temperature $t_0 = 16\ °C$, as well the movement of the grating with the respect to the annual movement of the system composed of the Sun and the Earth. At the same time, he abstracted from the influence of grating's movement because its speed was negligible when compared with the speed of light, as well as from the deviations of atmospheric pressure from its normal value (760 mm of height of the mercury column), because this variation had no sensible influence on the value of λ (1868, 14, 19). By this modification he obtained the relation $\log \lambda = \log \sin\varphi + \log \dfrac{e}{N} + \zeta$, where ζ stands for corrections brought in with respect to the orientation of the grating and the variation of temperature.

Utilizing this equation Ångström computed, given the measured values of e, φ, and ζ for both grating, the data for the spectra C, F of hydrogen. And then, based on all of the relevant data, he computed the *weighted arithmetic mean* values of the wave lengths of the spectral lines of solar light. For the four lines C, F, near G, and h – in the more modern notation H_α, H_β, H_γ, and H_δ – of hydrogen he obtained mean values, all in the range of 10^{-10} m.

These values were used by J. J. Balmer in the formulation of the first truly unifying formula for the wavelengths of the spectral lines of hydrogen. He was able to derive the following formula unifying the knowledge of the wave lengths of the four spectral lines H_α, H_β, H_γ, and H_δ (1885a; 1885b) (λ_0 stands for a constant):

$$\lambda = \lambda_0 \frac{m^2}{m^2 - 2^2} (m = 3, 4, 5, 6) \qquad (3.1)$$

and then generalized it into the following formula:

$$\lambda = \lambda_0 \frac{m^2}{m^2 - n^2} \qquad (3.2)$$

In the ensuing development of spectral analysis, formulas (3.1) and (3.2) were further generalized by J. R. Rydberg and W. Ritz. Rydberg in (1890a) brought in the idealization that the dispersion of light in air is negligible, that is, the light propagates with a speed identical to its speed in vacuum and employed the magnitude n labeled "wave number"[38] in the sense of "*the number of wave lengths in 1 centimeter*" (1890a, 13), that is, $n = 10^8 \cdot \lambda^{-1}$. This enabled him to restate Balmer's initial formula (3.1) into the form $n = n_0 \cdot \frac{m^2 - 2^2}{m^2}$ and thus $n = n_0 - \frac{4n_0}{m^2}$. Then he introduced the constant N_0 by means of the relation $N_0 = 4n_0$ and, so as $\lambda_0 = \frac{10^8}{n_0}$, he computed the value of the constant N_0. By means of this constant, Balmer's formula (3.1) for hydrogen turns into the sequence of the following formulas:

$$n_\alpha = N_0(\frac{1}{2^2} - \frac{1}{3^2}),\, n_\beta = N_0(\frac{1}{2^2} - \frac{1}{4^2}),\, n_\gamma = N_0(\frac{1}{2^2} - \frac{1}{5^2}) \qquad (3.3)$$

Early in 20th century, W. Ritz formulated (1908; 1911) a principle for series of spectra enabling to combine the already known formulas of spectral series of a substance, so that one obtains new formulas of other series of this substance, which can be used for the prediction of the existence of these other series. By applying this combination principle to (3.3) one obtains from these three formulas the following two additional formulas:

$$n_\beta - n_\alpha = N_0(\frac{1}{3^2} - \frac{1}{4^2}),\, n_\gamma - n_\alpha = N_0(\frac{1}{3^2} - \frac{1}{5^2})$$

Thus, one predicts here the existence of a spectral line for which should hold the formula

$$n = N_0(\frac{1}{3^2} - \frac{1}{m^2})\,(m = 4, 5) \qquad (3.4)$$

This prediction was confirmed by Paschen (1908), who – based on a series of experiments – proved the existence of a spectral series corresponding to the last

---

38 On the introduction of this magnitude see (Stoney 1872).

formula. This discovery was at the same time a confirmation of Balmer's universal formula (3.2) which in Rydberg's restatement is:

$$n = N_0 \left(\frac{1}{n^2} - \frac{1}{m^2} \right) \qquad (3.5)$$

Bohr In Part I of his article (1913) set as his aim to deal with "the mechanism of the binding of electrons by a positive nucleus" (1913b, 2). He conceptually approached this mechanism by means of Planck's view of radiation which he employed drawing on the "general acknowledgment of the inadequacy of classical electrodynamics in describing the behavior of system of atomic size" (1913b, 2). At the same time, Bohr clearly differentiated in his approach two distinct but still interrelated phases, *one*, in which he provided "a basis for a theory of the constitution of atoms" (1913b, 2), and *another one, based on the first one*, in which he gave a "an account … for the law of the line spectra of hydrogen" (1913b, 2).

As the object of his thought operations Bohr chose "a simple system consisting of a positively charged nucleus of very small dimensions and an electron describing closed orbits around it" (1913b, 3), and where this thought object was subjected to three idealizations: the mass of the electron is negligible when compared with that of the nucleus, the eccentricity of the elliptical orbit is zero, that is, the orbit is a circle, and the velocity of the electron is much smaller than the velocity of light.[39]

Bohr, by unifying classical mechanical description of the orbital movement of the electron (with mass m and charge e and where W is energy to remove it from the nucleus with charge E to infinity) with Planck's idea of a discrete energy being radiated by a vibrator with the frequency v, and where the energy radiated in one emission is $\tau h v$, where τ is positive whole number and h Planck's universal constant, obtained the relation

$$W = \frac{2\pi^2 m e^2 E^2}{\tau^2 h^2} \qquad (3.6)$$

Bohr interpreted this relation in such a way that for the respective values of τ one obtains the values for W which characterizes the configurations of atom when no energy is radiated by the atom. These configurations he viewed as characterizing the states of the atom; so he assigned to them the term "stationary states."

After Bohr dealt with the working of the mechanism of the hydrogen atom, he moved on to the derivation of the formula for the spectral lines of hydrogen. As

---

39 For detailed analysis of idealizations involved in the formulation of scientific laws see (Nowak 1972) and (Such 1978).

shown above, Rydberg's formula (3.5) holds for the spectral lines of hydrogen. So as Rydberg presupposed that it holds for the propagation of spectral lines in vacuum, it can be transformed by taking into account classical (Maxwellian) electrodynamics which views the radiation emitted by an atom as a *wave* propagating in vacuum with the speed c. For such a wave holds the relation $\lambda = c \cdot T$, where T is the period of its movement. The magnitude n is then transformed into $n = \dfrac{1}{c \cdot T}$, and by introducing the magnitude frequency υ as $\upsilon = \dfrac{1}{T}$ one obtains $\upsilon = c \cdot n^{40}$.

So, by multiplying the formula (3.4) by the magnitude c, one obtains

$$\upsilon = c \cdot N_0 (\frac{1}{n^2} - \frac{1}{m^2}) \tag{3.7}$$

Bohr then proceeded in the following way. He interpreted the experimental process of discharge in vacuum tube when the spectrum is experimentally produced as a process of moving the orbital electron to a great distance from the nucleus. This latter process he viewed as corresponding to the creation of a stationary state when the electron is bound by the nucleus. For the energy W radiated in this binding holds the relation (3.6), and so as the object of his thought derivations is still hydrogen, he put the absolute values of the charge of the orbiting electron and that of the nucleus as identical. He thus obtained from (3.6) the relation

$$W_\tau = \frac{2\pi^2 m e^4}{\tau^2 h^2} \tag{3.8}$$

Then, the energy produced when the atom passes from the state characterized by τ_1 to that characterized by τ_2 is as follows:

$$W_{\tau 2} - W_{\tau 1} = \frac{2\pi^2 m e^4}{h^2} (\frac{1}{\tau_2^2} - \frac{1}{\tau_1^2}) \tag{3.9}$$

This energy he put equal to $h\upsilon$, where υ is the frequency of radiation and obtained

$$\upsilon = \frac{2\pi^2 m e^4}{h^3} (\frac{1}{\tau_2^2} - \frac{1}{\tau_1^2}) \tag{3.10}$$

---

40 For details of this see (Sommerfeld 1922, 157; 1923, 154).

The expression in the brackets here was used by Bohr in the *explanation* of already known spectral series and in the *prediction* of the existence of spectral lines as yet undetected. By putting $\tau_2 = 2$ and for a varying τ_1 he *explained* Balmer's series and for $\tau_2 = 3$ he *explained* Paschen's series. He *predicted*, by putting $\tau_2 = 1$, or 4, or 5, the existence of spectral series both in the extreme ultra-violet and the extreme ultra-red, which "are not observed, but the existence of which may be expected" (1913b, 8). Already in 1914 the series corresponding to $\tau_2 = 1$ was detected; few years later the series for τ_2 equal to 4 and 5 were also experimentally detected.[41]

3.2.2 An Attempt at an Epistemological Generalization

From the reconstruction of the main features of the development of the experimental and theoretical dimensions of spectroscopy and its relation to Bohr's quantum theory of the hydrogen atom, I draw the following conclusions.

First, spectroscopy in the experiment performed by Ångström stands for the practical operations on the light coming from the Sun. *Second*, these operations draw on a concatenation of theories like geometry, theory on which is based the construction and employment of thermometers, that is, what can be tentatively called as "thermoscopy,"[42] and the wave theory of light. *Third*, Bohr's theory is concatenated with these theories by bringing in yet another three theories: classical electrodynamics, classical mechanics and Planck's quantum theory.

Fourth, all these theories are stated in the context of idealizations enabling to conceptualize certain processes under the supposition that certain disturbing factors are not at work (e.g., the idealization that the spectral lines are not dispersed in air, or the three idealizations introduced by Bohr). And if these factors are at work, then the respective theories have to be modified (e.g., when Ångström brought in the correction factor ζ into the equation for the wave-length).[43] These idealizations as well as their abolishment with their respective theories drive also the practico-experimental operations (e.g., when Ångström attached a thermometer to his experimental apparatus).

Fifth, the concatenation of theories can thus be viewed as an interconnection of *closed* conceptual systems because – due to the idealizations – certain concepts pertaining to the disturbing factors are deliberately left out. Once this closure is

---

41 On this see (Lyman 1914), (Brackett 1922) and (Pfund 1924).
42 On this theory see (Chang 2004) an (Sherry 2011).
43 On the thought operations involved in such a modification see (Nowak 1972) and (Such 1978).

schematically represented by a closed curve and the symbols for the respective magnitudes are placed into these curves, the concatenation of theories as given in my case study can be represented by the following figure:[44]

Figure 3.1: Concatenation of Bohr's quantum theory with the pre-quantum theories involved in spectroscopic theory and spectroscopic experimental practice

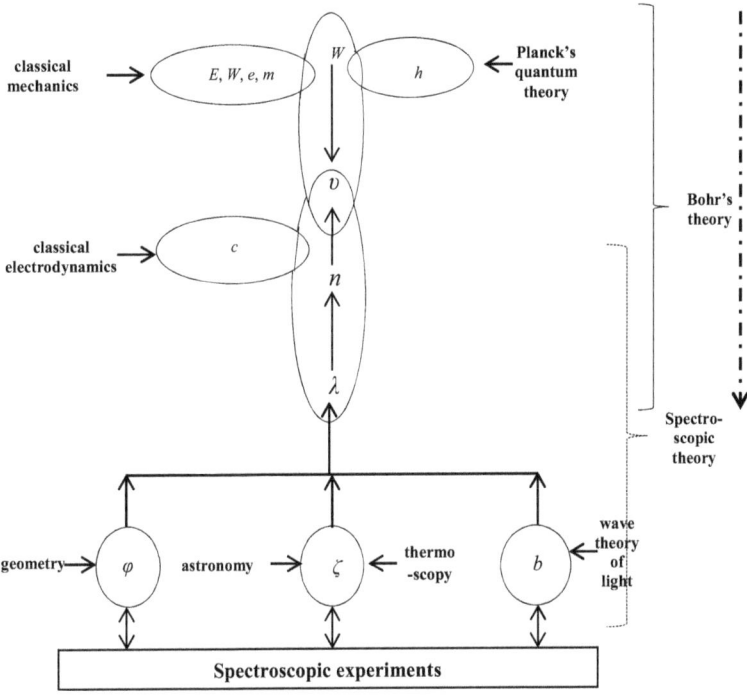

Sixth, the figure shows that the concatenation of theories is accomplished by certain magnitudes, for example, the magnitude v enabled Bohr to link the equation (3.8) for the magnitude W relation with the relation (3.7), which is the Rydberg formula modified by bringing in classical electrodynamics.

Seventh, contrary to the claim that one cannot test a theory because its consequences are theory-laden by this same theory, my case study shows that while the spectroscopic practico-experimental activity was loaded by a conceptual network set up a by a concatenation of several theories, still Bohr's theory acquired

---

44 Here I draw on (Pawson 1988); the term "closure" was introduced in (Bhaskar 1978).

its confirmation because it was connected to the results of experiments via a set of theories (geometry, thermoscopy, classical electrodynamics) which were stated before the formulation of Bohr's quantum theory, that is, they were of a *pre*-quantum nature.

Relying on these conclusions I can now evaluate the claims made by Feigl and Hempel. They were right in their claim that *only a network of theories can be subjected to experimental tests*. They were, however, *wrong in their claim that a theory can be tested by its observational consequences*. Bohr's theory is tested by theories concatenated with it, but which state nothing about observables, for example, the actual values of the wave-lengths of spectral lines of hydrogen which were computed on the basis of experiments were in the range of 10^{-10} m.

The concatenation of Bohr's theory with pre-quantum theories makes it possible to experimentally test this theory while at the same time to escape its own confines, because these pre-quantum theories and not Bohr's quantum theory were the basis of the testing experiments.

The involvement of experiments in Bohr's theory shows that natural objects are given to scientists not only in the mind as thought (e.g., idealized) objects which are dealt with by theories, but also in practico-experimental operation. So, *our sensual faculties are not only theoretical (theory-laden), but also practical (practice-laden)*.

Hempel was right in his claim that there exists, with respect to a certain theory, an antecedently given vocabulary which does not refer to observables. Curiously enough, Hempel completely bypassed the fact that *that vocabulary once restated in the light of that theory also does not refer to observables*, but for reasons which are different from reasons for the non-observability of entities referred to by the antecedently given vocabulary. The reason is that the *symbols and their meanings which are introduced for the first time in the theory, and which are not given in the antecedent vocabulary, "seep" into the explananda and predicanda derived from this theory* and so change the original meanings of symbols initially given in the antecedent vocabulary. So, e.g., the expression "$c \cdot N_0$" in (3.7) is restated in (3.10) as $\dfrac{2\pi^2 m e^4}{h^3}$, and so the formula for the phenomenon of spectral emission is completely tied to its explanans, the latter being Bohr's equation (2.8). The symbols "e," "h," and "m," together with their respective meanings appear in the explananda and predicanda because there are shifted here in the process of explanation and prediction from Bohr's explanans.[45] By this comparison one

---

45 For the sake of simplicity I use here only the term "explanans" as the basis for the thought derivation of both the explananda and predicanda; thus refraining from the use of the term "predicans."

finds out also that the interpretation of the symbols "n" and "m" in (3.1), (3.2), (3.5) and (3.7) is different from that of "τ_2" and "τ_1" in (3.10). While the former pair stands just for an ordered pair of numbers, the latter stands for the sequence of orbits of the orbiting electron. This transformation of the meaning of the vocabulary antecedently given into a consequently given one is represented in Figure 3.1 by means of the dashed arrow on the right side; it indicates the direction of derivation accomplished by Bohr and going from W via v "back" to the reinterpreted magnitudes n and λ.

From point of view of this transformation of meaning it becomes apparent that Feigl is wrong to claim that Balmer's formula – even after Bohr stated his theory of the hydrogen atom – stood for a *relative stable*, that is, a *not modified scientific law* (1974, 8–10). That such a transformation really takes place is readily seen when one compares the claims made by physicist after they performed the respective experiments. Paschen stated "The two lines are without doubt the first two terms ($m = 4$ an $m = 5$) of the series of hydrogen predicted by Mr. Ritz on the basis of his combination principle" (1908, 566). This means that his experiments were initially driven by the prediction based on the pre-quantum formula (3.5) and only later their results were *explained* in the process of theory construction by Bohr.

The situation is different with respect to those spectroscopists who performed their experiments with the knowledge about Bohr's theory.[46] So, for example, Brackett declared: "With a long hydrogen tube, viewed end on, as a source, lines have been observed at 4.05 μ and 2.61 μ, which, according to Bohr's theory, may be explained as due to an electron falling from the fifth to the fourth and from the sixth to the fourth rings respectively" (1922, 206).[47] Pfund declared in a similar manner (1924, 196):

> Upon the basis of Bohr's theory, a calculation was carried out for the wave-length emitted upon the falling back of an electron from the 6th to the 5th ring. The precise formula used was:
>
> $$\frac{1}{\lambda} = 109{,}677.7\left(\frac{1}{n_1^2} - \frac{1}{n_2^2}\right) \quad n_1 = 5, n_2 = 6$$
>
> And the resulting wave-length of this line came out as $\lambda = 7.46\mu$. In view of the fact that the observed maximum was found at 7.40μ—no great accuracy being claimed for wavelength measurements—it seems probable that the line found is the first member of the new series represented by the above formula.

---

46 Lyman did not relate his experiments to Bohr's theory but instead to his own experiments performed in the pre-quantum period; on this see (Lyman 1904; 1906).
47 Here μ stands for 10^{-6} m.

3.3 An Epistemological Way Out

Based on the above reconstruction I propose the following thesis. From the point of view of philosophy of science, it is essential to draw on the distinction between two fundamental and mutually irreducible, but at the same time interconnected ways by means of which humans appropriate the world of things outside the mind: a) by means of *thinking*, which stands for operations on the things of the mind and b) by means of *practical operations* on things in the world, which involve the employment of human bodies with their perceptual systems (i.e., I view the intertwining of thought operations as completely separated from sensory perception).

As a consequence of this approach I propose to abandon, in the framework of *philosophy of science*, the standard epistemological differentiation between perceptual and non-perceptual types of *knowledge*. I also reinterpret the categories *empirical knowledge* and *theoretical knowledge*, so that the former does not pertain to *sensually perceptibles* and the latter to *sensually non-perceptibles*. The designation of the empirical and the theoretical as two forms of *knowledge* makes sense only when both of these forms are viewed as the result of thought operations, and the former type is understood, when compared to the latter, as *more closely related to practical operations* in the world. So, in my approach, scientific *knowledge*, once it is understood as the result of thought operations on things of the mind, that is, as the result of the productive activity of thinking, should not be epistemologically investigated into by means of the category dichotomy "*observable vs. nonobservable*," nor by the category dichotomy "*the sensually perceptible vs. the sensually non-perceptible.*"

The intertwining of practical operations on things of nature with thought operations on things of the mind can be represented by means of the following figure:

Figure 3.2: The cycle of cognition unifying practico-bodily operations and thinking

Here "PB" stands for practico-bodily operations on things in nature, while "TH" stands for thinking in the sense of thought operations on things in mind. In order to symbolize that both the practical operations and the thought operation change, the figure has to be "broken up" and symbolized as follows:

$$\ldots \to PB \to TH \to PB' \to TH' \to \ldots$$

Let me first deal with the process of thinking; its cycle can be symbolically expressed as TH → PB→ TH'. I understand the term "thinking" as consisting of *operations creating and transforming things in the mind* (*thought things*) and at the same time as re*producing things outside mind* (*things of nature*) *as things in the mind*. The thought thing does not have character of a percept produced by the interaction of a physical entity with a perceptual organ but it is an *ideal thing standing for and at same time representing in the mind, the real thing*.

The ideal thing's standing for and representing a thing outside the mind enables one to think the real entity in its "pure" form (e.g., when Bohr produced his thought object which he labeled "hydrogen atom") by bringing in the above mentioned three idealizations pertaining to that thought object.

In my approach, a thing related *to* and its being *for* thinking is knowledge *about* and cognition *of* this thing. This knowledge and cognition passes through various stages characterized by Figure 2.2 in Chapter 2, and what gradually changes as thinking passes through these stages is the thought thing itself. So, thinking performs operations on itself: it is a unity of processes and of results of these processes. It reflects on its own, already-given content and posits it anew and transforms it by subjecting it to operations. This characteristic of thinking can be expressed as the development of the *intensional* content of thinking due to its *intentional* orientation of its processes on itself. At the same time, thinking – by means of its thought operations – appropriates the world of things by reproducing them in the mind, and so *serves in the discovery and determination of the objective features of the natural world*.

Let me now deal with the *practical operation* in the world of things; its cycle can be symbolically expressed as PB → TH → PB'. Here PB stands for the employment of human bodies which act, by employing things of nature so that they act on other things of nature. The latter things I label *objects of action* and the former as *the instruments of action*; what unifies these two classes of things is the practico-bodily intervention. By introducing the category of practical operations I clearly oppose the view that "the primary concern of science ... [is] the testing of hypothesis and the acquisition of knowledge" (Shapere 1982, 508). I believe that the *primary concern of natural science in the long run is the preservation and extension of the practical control by human beings of nature surrounding them.*

For a better understanding of the importance of these categories in the description of practico-bodily operations let me turn to the views of B. Van Fraassen on the relation of perception to practical operations. He gives, when explicating the difference between the categories *observing* and *observing that*, the following example: "Suppose one of the Stone Age people recently found in the Philippines is shown a tennis ball ... From his behavior, we see that he has noticed [it]; for example, he picks up the ball and throws it. But he has not seen that it is a ball ... for he does not have those concepts" (1980, 15).

While in Van Fraassen's view for a thing to be observed as a tennis ball, the observer must have the concepts (rules in mind) for playing tennis, in my approach, what is at stake is not only to be *observed* and to *have concepts of certain rules*, but also the *practical execution of the known rules*, their practical employment, where the latter involves the use of an instrument and of an object on which this instrument is being applied by the human who brings them together in interaction by applying his/her body. In the case of practical execution of the rules of tennis, the result of the interaction is a change of the state of affairs in the world: the object on which the instrument acts is shifted in space. Thus for a thing to be a tennis ball, a person – in addition to knowing the respective rules – must bodily act. A thing not used in a sequence of bodily operations standing for the execution of a set of known rules of the tennis game is not a tennis ball. If the Stone Age individual threw a thing, used before by another individual as a tennis ball in a tennis match, to kill an animal, then this thing is turned in the action of that Stone Age individual into a hunting instrument.

Let me now know turn to the following three claims made by Van Fraassen (1980, 16, 57–58, 81):

> [1] [T]he moons of the Jupiter can be seen through a telescope; but they can also be seen without a telescope if you are close enough. ... A look through a telescope at the moons of the Jupiter seems to me a clear case of observation, some astronauts will no doubt be able to see them as well as from close up. But the purported observation of micro-particles in a cloud chamber seems to me a clearly different case ... while the particle is detected by means of the cloud chamber, and the detection is based on observation, it is clearly not a case of the particle's being observed ...
>
> [2] I regard what is observable as a ... function of facts about us *qua* organisms in the world ...
>
> [3] [W]hat is the world in which I live, breathe and have my being, and which my ancestors of two centuries ago could not enter? It is the intensional correlate of the conceptual framework through which I perceive and conceive the world. But our conceptual framework changes—but the real world is the same world.

So, while Van Fraassen focuses on what is and is not observable, in my view the focus of philosophical analysis should be here on the *practical operations* in the world of things. Not only are *thought entities* (i.e., the meanings or intensions of our linguistic expressions) subject to changes, due to operations in the mind, *the world of things* itself (which human beings continually populate with new artifacts) is likewise subject to changes, due to practical operations.

Thus, from the point of *practical operations*, there is no difference *in principle* between what Van Fraassen labeled as "observability of the moons of the Jupiter" and the "non-observability of particles in a cloud chamber." In both cases human beings construct in the external world special equipment and thus change this world. In the former case, more and more precise telescopes, based on the use of light, ultra-sound, X-rays etc., are constructed, and even may be a spaceship will be produced in the future to travel to the moons of the Jupiter; in the latter case, bubble chambers, accelerators, etc., are being constructed. It holds thus, that from the point of view of *practical operations* there is no difference between human operations in the macro world and in the micro world.

While, according to Van Fraassen, what is and is not observable depends solely on our biological outfit, which is the result of our past biological evolution (say, of the genetic code) as a species, in my view observability *depends on our practical operation in the external natural world by means of our bodies/senses plus our experimental and industrial equipment. Our present senses are thus the result of the history of human practical operation, including experimentation and the construction of instruments and devices, in the external natural world*, that is, not only of our past *natural* history as a species, but also, and primarily, of our past *social* history. So, to the questions, *how many colors a human eye is able to distinguish* and *how many fragrances a human nose can distinguish*, the answer should be: *as many as humans are able to produce practically*.

That human beings operate both in outer space by placing into it artificial satellites and space stations, and in the micro world which they physically change, is best understood when one takes into account the changes in the structure of microprocessor and of the techniques used in their production. The first computer used by the author of this book was equipped with a Z80 processor with "only" 8500 transistors placed on an area of 18 mm². The next one was equipped with an 80286 processor with $134 \cdot 10^3$ transistors on an area of 49 mm². Finally, the computer on which this book was written employs a Core i3 processor with $382 \cdot 10^6$ transistors on an area of 81 mm². The success in placing an ever increasing number of transistors on the same unit of area of the processor is based on the ability to operate practically in the micro world by employing computer driven machines performing operations on things in the regions of ever decreasing dimensions: starting from $4 \cdot 10^{-6}$ m in the production of the Z80 processor and ending in the production of the Core i3 in regions with the dimensions of $45 \cdot 10^{-9}$ m.

The conclusion I can draw here is the same as the one arrived at in the case of the experiments leading to the production of the linear spectra of hydrogen. Processes in regions with dimensions in the range of 10^{-10} m in which the spectroscopists like Ångström were operating when manipulating spectral lines, like processes in regions with dimensions in the range of 10^{-9} m used now in the production of processors are *not perceptible* but are *practically* triggered, controlled, manipulated and employed, and they are also *thought* by means of concatenation of the respective scientific theories. Thus, what is relevant for human beings *as* human beings is not that "what is observable is a function of facts about us *qua* organisms in the world," as claimed by Van Fraassen, but their ability to operate both practico-bodily in and on nature and in the mind on thought objects of nature.

After dealing separately with the cycle TH → PB → TH' as the cycle of thought operations yielding new thought things and the cycle PB → TH → PB' as the cycle of practical operations in the world of things changing the state of affairs in this world by producing new things in it, let me now consider them in their unity. The mutual interactions of things in the natural world are transformed into human practico-bodily operations with things once these operations are driven by the productive activity of thinking. When the triggering and renewal of practico-bodily operations is mediated by thought operations and draws on the results of the latter, then practico-bodily operations can implement the new results produced in thinking by transforming the things of nature in new ways by employing other things on them. These practical operations yield thus new, previously not given state of affairs which can in turn be thought and thus drive thinking to new thought objects as its results.

The fact that human beings are the locus of practical operations going into the external natural world and of thinking related to both the thought world and, via communication, to the thinking of other human beings, I express by the following figure:[48]

---

48 Here I draw partially on (Habermas 1981).

Figure 3.3: Practical operations and thought operations performed by mutually communicating and practically cooperating human beings

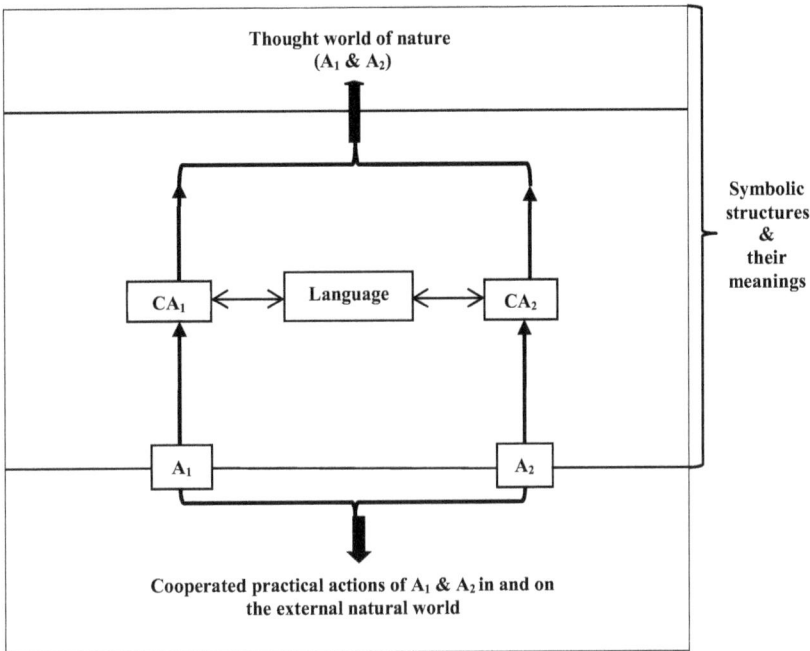

Here the term "Thought world of nature" stands for the state of affairs intersubjectively thought by actors A_1 and A_2. By means of statements integrated into communicative acts (CA_1, CA_2) performed by A_1 and A_2 these thought state of affairs become the objects of mutual communication between A_1 and A_2. One can speak here about a relation to the thought world indicated by the arrow going upwards. The term "external natural world" given at the bottom of the scheme, stands for the state of affairs into which the actors intervene by employing means of action on the objects of action, and where these means and objects are unified by performing bodily operations. The arrow here indicates the established practical relation to the world of nature.

99

Chapter 4: Kantian and Post-Kantian Themes in Early Quantum Mechanics

The aim of this chapter is to deal with some Kantian themes in early quantum mechanics, where by these themes I understand certain aspects of Kant's epistemology presented primarily in the *Critique of Pure Reason* which reappeared in the course of the formulation and discussion in early quantum mechanics in the second half of 1920's. At the same time I shall try to show that an adequate epistemological treatment of early quantum mechanics requires a further development and change of the framework and content of Kant's epistemology.

My approach to early quantum mechanics is based on the view that one can discern in a natural-science discipline an *experimento-conceptual* dimension, that is, the dimension where it practically encounters nature and where it tries to treat certain problems which it faces when intervening into nature by treating them in conceptual systems. In addition to this dimension, I identify two more: a *meta-conceptual* dimension, where the choice of certain concepts in the first dimension is subjected to a specific reflective endeavor, and a *methodological* dimension, where issues like methods of explanation, and prediction employed in the conceptual dimension are subjected to a special analysis. With respect to the meta-conceptual dimension I will try to show that Kantian themes like intuitiveness (*Anschaulichkeit*), observability, the existence of a thing-in-itself, the applicability/nonapplicability of space and time as forms of intuition loomed large in the 1920's.

In my approach I shall unify the epistemological analysis of the meta-conceptual dimension of early quantum mechanics with a summary of Kant's epistemology and with an analysis of the experimento-conceptual dimension of early quantum mechanics. This unification shall enable me to improve upon the approach of the more recent philosophy-of-science literature to early quantum mechanics which provides a descriptive chronology of the epistemological claims made by the creators of quantum mechanics, but without confronting these statements with the theories they produced in order to find out if these epistemological statements really hold for these theories. So, for example, M. Beller declares that Heisenberg presented in his *Umdeutung* paper (1925) an "extreme Machian approach which eliminated everything but immediate perception (intensities, frequencies, and polarization)" (Beller 1999, 30), but she does not confront this claim with an analysis of the matrix mechanics, in order to find out if the radiation-magnitudes (*Strahlungsgrössen*) appearing as elements of the matrices really pertain to something perceptible.[49]

Although I cannot deal in detail here with the methodological dimension of early quantum mechanics, in the conclusion of the chapter I will delineate the framework and direction in which this dimension should be dealt with in the future.

All translations of the German texts are mine. In quotations from Kant I draw on the English translation (Kant 1998).

4.1 Kant on Intuition, Appearance, the Thing-in-Itself, and Categories

Kant's division of the main part of the *Critique of Pure Reason* (*CPR*) into *Transcendental Aesthetics* and *Transcendental Logic*, the latter being subdivided into *Transcendental Analytics* and *Transcendental Dialectics*, is based on his view that there exist two sources of human cognition (*Erkenntnis*) (A50=B74; 1998, 193):

> Our cognition arises from two fundamental sources in the mind, the first of which is the reception of representations (the receptivity of impressions), the second the faculty for cognizing an object by means of these representations (spontaneity of concepts); through the former an object is given to us, through the latter it is *thought* in relation to that representation ... Intuition and concepts therefore constitute the elements of all our cognition.

In addition, Kant differentiates "levels" of cognition. "All our cognition starts from the senses, goes from there to the understanding, and ends with reason, beyond which there is nothing higher to be found in us to work on the matter of intuition and bring it under the highest unity of thinking" (A298=B355; 1998, 387).

Let me now turn to the Transcendental Aesthetics section of the *CPR* where he states the following (A20=B34-35; 1998, 155-156):

> The effect of an object on the capacity for representation, insofar as we are affected by it, is *sensation*. That intuition which is related to the object through sensation is called *empirical*. The undetermined object of an empirical intuition is called appearance. I call that in the appearance which corresponds to sensation its *matter*, but that which allows the manifold of appearance to be intuited as ordered in certain relations I call the *form* of appearance.

Then he performs a two-step thought operation. First, he isolates "sensibility by separating off everything that the understanding thinks through its concepts,

---

49 For a similar approach see (Camillieri 2009).

so that nothing but empirical intuition remains" (A22=B36; 1998, 157). Next, he isolates from empirical intuition "everything that belongs to sensation, so that nothing remains except pure intuition and the mere form of appearances" (A22=B36; 1998, 157).

Kant views *space* and *time* as forms of intuition, which do not have their source in the external experience of appearances (A22-23=B37-38, A31-32=B46-47; 1998, 157, 162). As will be show below, Kant's concept *Anschaung* reappears in the epistemological discussions on the new quantum mechanics. Another important concept here is *Erfahrung* (i.e., *experience*), which he characterizes as follows (B1; 1998, 136):

> There is no doubt whatever that all our cognition begins with experience; for how else should the cognitive faculty be awakened into exercise if not through objects that stimulate our senses and in part themselves produce representations, in part bringing the activity of our understanding into motion to compare these, to connect or separate them, and thus to work up the raw material of sensible impressions into a cognition of objects that is called experience? *As far as time is concerned*, then, no cognition in us precedes experience, and with experience every cognition begins.

From the treatment of perceptions and intuitions in the *Transcendental Aesthetics* Kant advances to the treatment of understanding (*Verstand*) in the *Transcendental Analytics* by making the following claiming (A51=B75; 1998, 193):

> If we will call the *receptivity* of our mind to receive representations insofar as it is affected in some way *sensibility*, then the faculty for bringing forth representations itself, or the *spontaneity* of cognition, is the *understanding* ... the faculty for thinking the object of sensible intuition.

Accordingly, as long as they are related to experience, the elements of pure understanding are subject-matters of transcendental analytics are (A62=B87; 1998, 199). For Kant, the basic concepts of understanding which he labels as *categories* are central to his position.

The Kantian categories display a twofold, so to say, Janus faced, dimension. On the one hand, they are not grounded in the objects with which humans interact via perception, but, on the other hand, they enable one to cognize (*erkennen*) these things by thinking them. Due to the latter feature, they are related to and not separable from sense-objects. That is, "categories ... can have a determinate significance and relation to any object only by means of the general *sensible condition*" (A245; 1998, 344) and these can be related only "to object of the senses" (A246=B303; 1998, 345). The Janus-faced nature of Kant's categories can be expressed by the following pair of predicates: *innerworldly - outerworldly*. Categories of cognition have, due to their *a priori* nature, an *outerworldly* status. But, because "we cannot *think* any object except through categories" (B165; 1998, 264), they enable, once unified with intuition, us to cognize (*erkennen*) objects existing in the world, thereby fulfilling an *innerworldly* function in cognition.

At the same time Kant excludes the possibility that categories can be used beyond the framework of objects given in perception. This limitation, on Kant's part, of the employment of the concepts of understanding then delineates his approach to the *Transcendental Dialectics* which deals with the characteristics of cognition at the level of reason (*Vernunft*), namely, the employment of concepts beyond the perceptual realm (A62=B87–88; 1998, 199), to which *no object in sensible experience* corresponds. Such non-perceptual objects are God, mind and world (as a whole). The aim of *Transcendental Dialectics* is to bring to the fore that in the employment of concepts beyond the realm of experience (A295=B352; 1998, 385)

> we have to do only with *transcendental illusion*, which influences principles whose use is not ever meant for experience ... but which instead ... carries us away beyond the empirical use of the categories, and holds out to us the semblance of extending the *pure understanding*.

Based on my analysis of the early quantum mechanics, I will try to restate the subject matter of the *Transcendental Dialectics* and thereby reinterpret the Kantian meaning of the term "experience" (*Erfahrung*). As a preparation for this restatement and reinterpretation, let me now deal with Kant's approach to the pair of epistemological terms: *appearance – thing in itself*.

Kant declares with respect to the human capability to cognize *a priori* that "we can never get beyond the boundaries of possible experience ... that such cognition reaches appearances only, leaving the thing in itself as something real but uncognized by us" (Bixx-xx; 1998, 112). He justifies this claim by stating the following: "... that which necessarily drives us to go beyond the boundaries of experience and all appearances is the *unconditioned*" (Bxx; 1998, 112), where the latter "must not be present in things insofar as we are acquainted with them, as things in themselves (insofar as they are given to us), but rather in things insofar as we are not acquainted with them" (Bxx; 1998, 112).

The argument for the nonknowability and noncognizability of the thing in itself I interpret – and *accept* – in the sense that we cannot know the thing in itself as long as it exists in itself (i.e., is just *a thing outside cognition about which we know only that it is, but nothing else*). But, at least in my view, Kant has also yet another understanding of the term *thing in itself*. He understands it as the *ground of the appearance*, declaring first that "we can have cognition of no object as a thing in itself, but only insofar as it is an object of sensible intuition, i.e. as an appearance;" (Bxxv–xxvi; 1998, 115), and then adds (Bxxvi–xxvii; 1998, 115):

> Yet the reservation must also be well noted, that even if we cannot cognize these same objects as things in themselves, we at least must be able to think them as things in themselves. For otherwise there would follow the absurd proposition that there is an appearance without anything that appears.

Thus, according to Kant, *while at the level of appearances there is an interlinking of knowing and thinking, this interlinking is not realizable with respect to the ground of the appearances*.[50] Kant declares the following: "what the things may be in themselves I do not know, and also do not need to know, since a thing can never come before me except in appearance" (A277=B333; 1998, 375).

He expresses this negative delineation of the term *thing in itself* by assigning to it the status of a *mere* thing in thought; "The thing in itself = x is a mere thought-thing *ens rationis ratiocinantis*" (1938, 421; 1993, 184). Below I will, with respect to Schrödinger's ψ-function, try to show that a *Gedankending* can be understood in a positive way.

4.2 The Early Matrix Mechanics and Wave Mechanics

The creation of the new quantum theory both in its matrix and wave form can be viewed as the result of the intersection of two developments taking place inside the "old" quantum theory.[51] One was the realization that the results of the Bohr-Sommerfeld theory failed "empirically" with respect to the behavior of atoms more complex than the atom of hydrogen (e.g., the ionization potential of helium computed on the basis of this theory did not fit the potential computed on the basis of the results of experiments).[52] Another problem which surfaced with respect to this element was the description of quantum orbits in the case of excitation should yielded characteristic of its spectra which were not experimentally confirmed.[53] And the Bohr-Sommerfeld model also failed to come to terms with the behavior of the spectra of atoms once the latter were subjected to the action of a combined external electrical and magnetic field.

---

50 Here I draw on (Rescher 1981).
51 For a detailed factual description of the creation of quantum mechanics see, for example, (Beller 1999), (Camillieri 2009) and (Darrigol 1992).
52 On this see (Kramers 1923) and (Van Vleck 1922; 1923).
53 On this see (Born and Heisenberg 1923).

The second development, directly linked to the first, was that the physicist in the early 1920's viewed those failures, to use the term of O. Darrigol (1992, 181), as *path killers*, namely, that they realized that the revision of existing quantum theory has to performed in its very conceptual basis, namely, at the level of kinematics. So, W. Pauli writes in a letter to N. Bohr from February 21, 1924 as follows: "As the most important question appears to me: to what extent one can speak at all about orbits of the electrons in the stationary states. I think that this should not be taken for granted" (Pauli 1979, 148) and on December 12 of the same year he writes to Bohr "the relativistic formula for doublets appears to show me without doubt that … the kinematic concept of motion of the classical theory has to undergo profound modifications" (Pauli 1979, 188–189).

Heisenberg presents a similar line of reasoning in a letter to Pauli of July 9, 1925 when he declares that "my endeavor goes into the direction of killing the concept of orbits, which cannot be observed, and to replace it by a suitable one" (Pauli 1979, 231)

This statement reappears in the abstract of Heisenberg's *Umdeutung* paper (1925) in the following form: "In this work an attempt is to be made to obtain foundations for a quantum-theoretical mechanics, which is based exclusively upon relations between magnitudes which are observable in principle" (1925, 879). And in the body of the text he delineates his points of departure in a more detailed way as follows (1925, 879):

> It is well known that against the formal rules which in general are used in quantum theory for the computation of observable magnitudes (for example, of the energy of the hydrogen atom) it is possible to raise the serious objection, that these rules contain as an essential element relations between magnitudes which apparently are not observable in principle (as, for example, position, period of the electron), that is, those rules lack any perceptible (*anschaulich*) physical foundation.

As an alternative strategy to this deficit he proposes "to try … to form a quantum-theoretical mechanics, in which occur only relations between observable magnitudes" (1925, 879).

Next Heisenberg considers oscillations of the classical coordinate $x(n, t)$ assigned to an electron in an atom. These oscillations can be understood as superpositions of harmonic oscillations, thus allowing Fourier's theorem to be applied. For periodic and nonperiodic motions holds

$$x(n,t) = \sum_{\alpha=-\infty}^{+\infty} A_\alpha(n) e^{\alpha(2\pi i \nu(n)t)}, \quad x(n,t) = \int_{-\infty}^{\infty} A_\alpha(n) e^{\alpha(2\pi i \nu(n)t)} \qquad (4.1)$$

These oscillations are, however, not assigned to certain orbits of the electron in the atom, but are associated with the transition between two stationary states n and $(n - \alpha)$ which lead to the emission of the radiation. This radiation is characterized by means of three magnitudes: the amplitude A_α, the frequency v, and the intensity proportional to $A_\alpha(n) \cdot v^4$. The amplitude is related to the probability of a spontaneous passage from the stationary state n to that of $(n - \alpha)$, while for frequency holds $v(n, n - \alpha) = \dfrac{1}{h}\{W(n) - W(n - \alpha)\}$ as well as combination relations based on Ritz's principle dealt with in Chapter 2 as well with below.

The quantum theoretical magnitude associated with $x(n, t)$ is viewed as an *aggregate* of the magnitudes $A(n, n - \alpha)e^{2\pi i v(n, n - \alpha)t}$ or, as latter found out by Born, a *matrix* which can be symbolized as $(A(n, n - \alpha)e^{2\pi i v(n, n - \alpha)t})$ and stands for x as the quantum mechanical counterpart of the classical coordinate $x(n, t)$. The rest of the story, namely, the assignment of matrices in the papers (Born – Jordan 1925; Born – Heisenberg – Jordan 1925) – with their specific algebra – to the classical physical magnitudes is well known and need not be repeated here.[54]

With respect to the *Umdeutung* paper it is worth noting that Heisenberg's claim about the involvement in his new quantum mechanics of only those magnitudes which are observable was accepted by other physicist in the years to come. So, Sommerfeld declares (1927, 231–232):[55]

> Heisenberg starts from the epistemological principle that the physical description of phenomena should use only observable elements. In the atom these are frequencies of the spectral lines and their intensities ... All the other mechanical magnitudes, for example, the position of the electron on the orbit, the period of the revolution are not observable and drop out of the theory.

Let me now turn to wave mechanics as presented in the *Mitteilungen* I through IV (1926a; 1926b; 1926d; 1926e) and in the paper (1926c) dealing with the relation of wave mechanics to matrix mechanics.

Schrödinger introduces in the first *Mitteilung* the Hamiltonian partial differential equation $H(q_i, \dfrac{\partial W}{\partial q_i}) - E = 0$, into which he substitutes for action-variable S a new variable ψ so that it holds $S = K\log\psi$, where K has the dimension of action. He obtains the equation $H(q_i, \dfrac{K}{\psi}\dfrac{\partial \psi}{\partial q_i}) - E = 0$ and imposes on it the variation principle. He chooses as the object of reflection the hydrogen atom subjected to the two idealizations: the speed v of its electron with charge e and mass m is

---

54 For a detailed account of this see (Born 1926), (Birtwistle 1928) and (Darrigol 1992).
55 A similar claim is made in (Bohr 1925; 1926).

much smaller the speed of light c, that is, $\frac{v}{c} = 0$, and the atom is not disturbed by external forces, $Fe = 0$. He puts $K = \frac{h}{2\pi}$ and arrives at the wave equation of an one-electron atom (1926a, 362)

$$\Delta\psi + \frac{2m}{K^2}(E + \frac{e^2}{r})\psi = 0 \tag{4.2}$$

From this equation he is able to prove that for $E < 0$ the variation problem has a solution only if it holds

$$-E_l \frac{me^4}{2K^2 l^2} \quad (l = 1, 2, 3, \ldots) \tag{4.3}$$

This relation is the quantum-theoretical reinterpretation of Balmer's formula for the spectral terms of hydrogen in its quantum form dealt with in Chapter 2.

In the second *Mitteilung*, an optico-mechanical justification for the wave-theoretical approach in quantum mechanics is given by Schrödinger who declares (1926b, 506, 508):

> In order to form an image of the manifold of the possible processes, we must proceed from the *wave equation* and not from the fundamental equations of mechanics ... the true laws of quantum mechanics do not consist of certain rules for the individual paths, but in these true laws the elements of the whole manifold of paths of a system are bound together by means of equations ... All these assertions are going ... systematically in the direction of relinquishing of the concepts "place of the electron" and "path of the electron". If one does not decide in favor of this relinquishing, these assertions remain contradictory.

At the same time he relates this contradiction to Heisenberg's attempt to get rid of the concepts of space and time by characterizing them via the term "form of thinking" (1926b, 508–509):

> This contradiction has been so strongly perceived that it has been doubted whether what goes on in the atom could be at all be integrated into the space-time form of thinking. From the philosophical standpoint, I would consider a conclusive decision in this sense as equivalent to a complete surrender.

This negative evaluation on his part he justifies in the following Kantian manner: "For we cannot really change the forms of thinking, and what we cannot understand within them we cannot understand at all" (1926b, 509).

All the remaining forms of the wave equation were dealt with in Chapter 1.

As a consequence of the above-mentioned point of view on the role of the concepts of space and time as forms of thinking, Schrödinger states in another paper, that Heisenberg's matrix mechanics lacks "intuitiveness" (1926c, 734). What surfaces here is an important difference between Schrödinger's and Heisenberg's understandings of the German term "anschaulich." The former views the concepts of space and time as forms of thinking which enables one to *understand what is perceived* (i.e., they *enable observation*). On the contrary, the latter, in the *Umdeutung* paper, views a rule for the computation of a physical magnitude (for example, energy) as not *anschaulich*, once this rule involves magnitudes (e.g., space and time) which are not *observable*. This means that in Heisenberg's approach observation is *not* enabled by the concepts of space and time; he reduces the *observable* to the *perceivable*. If this claim really holds, Heisenberg's employment of the term "anschaulich" in the *Umdeutung* paper has to be translated as "perceptible" while its understanding by Schrödinger's in the papers (1926b; 1926c) has to be translated – following the standard translations of the *Kritik der reinen Vernunft* (1973; 1998) – as "intuitive."

The difference between their respective approaches to the term "anschaulich" is then placed by Schrödinger into a broader epistemological context where he compares wave mechanics with matrix mechanics. He emphasizes that while from the mathematical point of view they are equivalent, from the physical point of view they are not. This differentiation between these two points of view he explicitly puts into opposition to Mach's epistemological approach to physical theories, according to which the aim of physical theories should be to give a "description of the empirical connections between observable magnitudes, that is, a description which reproduces the connection, as far as possible, without the mediation of elements which are not observable in principle" (1926c, 751).

The issue of an anti-Machian interpretation of wave mechanics was also explicitly stated by Sommerfeld who, when comparing the matrix and wave mechanics, declared: "While Heisenberg cautiously sticks to experience, Schrödinger goes ... boldly beyond the experience and constructs wave-images, which stand behind the experience, thus are metaphysical (*metaphysikalisch*)" (1927, 232).

While Sommerfeld in (1927) did not explicitly subsume Heisenberg's approach in the *Umdeutung* paper under a philosophical position, both C. Lanczos and Einstein did. The former declared that if "all facts are of such nature that they yield only the coefficients of the matrices, then the matrix-like representation deserves preference (at least from the positivist point of view!), because it does not bring into the description of facts any principally unreachable element" (Lanczos 1926, 820).

Einstein, at least according to Heisenberg's recollections, also viewed the requirement of observability as stated in the *Umdeutung* paper as Machian in its very nature. In their discussion Heisenberg "confessed" to the Machian roots of

(1925): "The idea that a good theory is no more than a condensation of observation in accordance with the principle of thought economy ... goes back to the physicist and philosopher Mach" (Heisenberg 1969, 93). Einstein reacted as follows: "on principle, it is completely wrong to try founding a theory on observable magnitudes alone ... Mach's concept of thought economy ... strikes me as being just a bit too trivial ... if we wanted to speak of nothing but sense impressions, we should have to rid ourselves of our language and thinking" (Heisenberg 1969, 92–95).

4.3 The Myth of Observability: Kantian Themes in Early Quantum Mechanics

The view that physical theories should state relations between magnitudes referring only to observables is not the product of the crisis of the "old" quantum theory, but had been accepted by physicists already before that crisis. So, for example, as early as 1919 Pauli declared in a paper dealing with views of H. Weyl: "One would like to stick to the introduction only of principally observable magnitudes" (1919, 749–750).

The Machian background and basis of this claim is readily seen in P. Frank's paper of 1917, where he presents Mach's understanding of the tasks and possible aims of the exact natural sciences: "according to Mach ... physics is nothing else just a collection of statements about the connection of sensations of senses (*Sinnesempfindungen*) and the theories nothing else than economical expressions for the summary of these connections" (1917, 65). This epistemological approach I will label *as sensualism in a Machian cloth*, or *sensualism*, for short.

Ten years later, the sensualistic tenet reappears in Frank's epistemological evaluation of quantum mechanics, namely, that with respect to the latter (1928, 125):

> The meaning of the positivist conception ... is that all its statements can be resolved in statements about sensations, and that all others, such as atoms, electrons, etc., are only auxiliary concepts ... Nothing show as much as Heisenberg's quantum mechanics, that "intuitivity" may be sought only in the relation of the physical sentences to the experiences.

And few years later he states in the journal *Erkenntnis* (Heisenberg et al. 1931, 185–186):

> The bond between the observations is established by means of a nonobservable mathematical scheme, namely, the wave function. It is an auxiliary means by which the bond is established. The bond itself is not observable. The wave function is no reality. As a fictional bond this theory can be formulated ... The things that are being constituted out of the perceptions do not correspond to any objective reality existing outside the perceptions.

In order to scrutinize these sensualistic epistemological claims stated with respect to early quantum mechanics, let me return to Heisenberg's claim about the observability of the elements unified in the matrices of quantum mechanics. He states (1926, 685):

> The basic experience which is necessary for the derivation of the kinematic laws of the quantum theory is given by the Rydberg-Ritz combination principle. It states that each frequency $v(n\ m)$ of the observed spectrum can be represented as difference of two terms and, conversely, that each difference of two terms from the totality of the terms can appear as frequency of an observed spectral line, that is, it holds
>
> $$v(n\ m) = T_n - T_m, \quad v(n\ k) + v(k\ m) = v(n\ m)$$
>
> From the optical phenomena one can make the general conclusion that to the principally observable magnitudes belong the frequency and the amplitude and phase associated with it.

Therefore he labels both these equations as a "law of experience" (1926, 686).

Let me now try to demonstrate, by analyzing the introductory statement from his *Umdeutung* paper quoted earlier, that matrix mechanics is not related to observable magnitudes.

First, Heisenberg initially claims that in the "old" quantum theory there are "rules which ... are used for the computation of observable magnitudes" (1925, 879); that is, in this theory there are magnitudes which are *computed* by means of a *rule and* which, he claims, are at the same time *observable*. These two characteristics, however, can be achieved neither from the point of view of a *magnitude* of a physical theory, nor from the point of view of the *actual values* of such a magnitude. Once the actual values of magnitude p are computed on the basis a theory T by means of a rule R and the actual values of magnitudes $q, r, \ldots$ entering into R, then the *magnitude p* itself is in T introduced (defined) by means of R and the latter's application to *magnitudes $q, r, \ldots$*. But then, even if the latter magnitudes would be observable, p as a magnitude could not be an observable entity because R is not observable. And second, once the *actual values* of p are computed by the application of R to the actual values of $q, r, \ldots$, then, even if these values would be observable, the *actual values* of the magnitude p would be computed and not observed.

Second, a similar type of argument also can be applied also to Heisenberg's claim that the "old" quantum theory is deficient because it computes the *observable* magnitudes by means of *nonobservable* magnitudes. When the *magnitude p* is introduced (defined) in T by means of R and its application to *nonobservable magnitudes $q, r, \ldots$*, then p is not observable in T as a magnitude. And, once the *actual values* of p are computed by the application of R to the nonobservable *actual values* of $q, r, \ldots$, then the actual values of the magnitude p are not observable.

Third, in order to find out whether the radiation-magnitudes given in the matrices really refer to observable, one has to realize that a central element is here the magnitude υ standing for frequency. Its nonobservable nature becomes apparent if one takes into account how it was introduced in the prequantum spectral analysis in the 19th and early 20th centuries.[56]

In 1868 Ångström published the results of measurement of the wave lengths of lines in the solar spectrum. His aim was to produce a detailed atlas of the wavelengths of spectral lines given in the solar light which drew on experiments with optical gratings. As shown in Chapter 2, the equation used by Ångström for the computation of the wave length λ of the particular lines spectrum of the N-th order was $\log \lambda = \log \sin\varphi + \log \frac{e}{N} + \zeta$.

Even if I dealt with this period in the history of physics, I repeat here the most important formulas derived already in Chapter 2. Utilizing this equation Ångström computed, given the measured values of e, φ, and ζ for both grating, the data for the spectra C, F of hydrogen. And then, based on all of the relevant data, he computed the *weighted arithmetic mean* values of the wave lengths of the spectral lines of solar light.

These values were used in 1885 by J. J. Balmer in his formulation of the first truly unifying formula for the wave lengths of the spectral lines of hydrogen. He was able to derive the following formula unifying the knowledge of the wave lengths of the four spectral lines H_α, H_β, H_γ, and H_δ (1885a; 1885b) (λ_0 stands for a constant):

$$\lambda = \lambda_0 \frac{m^2}{m^2 - 2^2} \quad (m = 3, 4, 5, 6) \tag{4.4}$$

and then generalized it into the following formula:

$$\lambda = \lambda_0 \frac{m^2}{m^2 - n^2} \tag{4.5}$$

In the ensuing development of spectral analysis, formulas (4.4) and (4.5) were generalized by J. R. Rydberg and W. Ritz. Rydberg in (1890) – by employing the

---

56 Here I draw on (Ångström 1868), (Stoney 1872), (Balmer 1885a; 1885b), (Rydberg 1890), (Landauer 1898), (Ritz 1908; 1911) and (Baly 1912).

magnitude n labeled "wave number"[57] in the sense of "*the number of wave lengths in 1 centimeter*" (1890a, 13), that is, $n = 10^8 \cdot \lambda^{-1}$ – restated Balmer's initial formula (4.4) into the form $n = n_0 \cdot \dfrac{m^2 - 2^2}{m^2}$ and thus $n = n_0 - \dfrac{4n_0}{m^2}$. Then he introduced the constant N_0 by means of the relation $N_0 = 4n_0$ and, so as $\lambda_0 = \dfrac{10^8}{n_0}$, he computed the value of the constant N_0. By means of this constant Balmer's formula (4.4) for hydrogen turns into the sequence of the following formulas:

$$n_\alpha = N_0(\frac{1}{2^2} - \frac{1}{3^2}),\ n_\beta = N_0(\frac{1}{2^2} - \frac{1}{4^2}),\ n_\gamma = N_0(\frac{1}{2^2} - \frac{1}{5^2}) \qquad (4.6)$$

Early 20th century, W. Ritz formulated (1908; 1911) a principle for series of spectra enabling to combine the already known formulas of spectral series of a substance, so that one obtains new formulas of other series of this substance, which can be used for the prediction of the existence of these other series. By applying this combination principle to (4.6) one obtains from these three formulas the following two additional formulas:

$$n_\beta - n_\alpha = N_0(\frac{1}{3^2} - \frac{1}{4^2}), \quad n_\gamma - n_\alpha = N_0(\frac{1}{3^2} - \frac{1}{5^2})$$

Thus, one predicts here the existence of a spectral line for which should hold the formula

$$n = N_0(\frac{1}{3^2} - \frac{1}{m^2}), \quad (m = 4, 5)$$

This prediction was confirmed by Paschen (1908), who – based on a series of experiments – proved the existence of a spectral series corresponding to the last formula. This discovery was at the same time a confirmation of Balmer's universal formula (4.5) which in Rydberg's restatement is:

$$n = N_0(\frac{1}{n^2} - \frac{1}{m^2}) \qquad (4.7)$$

Formula (4.7) can be transformed once we take into account classical (Maxwellian) electrodynamics which views the radiation emitted by an atom as a *wave* with the speed of light c. For such a wave holds the relation $\lambda = c \cdot T$, where T is the period

---

[57] On the introduction of this magnitude see (Stoney 1872).

of its movement, and the magnitude n is then transformed into $n = \dfrac{1}{c \cdot T}$, and by introducing the magnitude frequency v as $v = \dfrac{1}{T}$ one obtains $v = c \cdot n$.[58] So, by multiplying the formula (4.7) by the magnitude c, one obtains

$$v = c \cdot N_0 \left(\dfrac{1}{n^2} - \dfrac{1}{m^2} \right) \tag{4.8}$$

This equation stands for the Rydberg-Ritz combination principle in its *prequantum-theoretical* form and corresponds to Heisenberg's reference to this principle in the quote I gave above.

What epistemological lessons can be drawn from this overview of spectral analysis? *First*, frequency v as a *magnitude* which appears in Heisenberg's matrices is *not an observable* entity because it is understood on the basis of rules which belong to two physical theories. The equation used by Ångström for the computation of λ is part of the wave theory of light and, because the magnitude c enters into formula for magnitude v, Maxwell's electrodynamics also has to be involved.[59] Thus, the magnitude of intensity of radiation proportional to v^4 which according to Heisenberg should be observable, in fact is not.

Second, the overview of the introduction of the magnitudes like n and v given above indicates that it stands for the *construction* of a theory of spectral analysis, where this construction takes place by the *interchaining* of the wave theory of light and of Maxwell's electrodynamics. To such a two-component chain Heisenberg hooks up his quantum matrix theory by bringing in for the magnitude v given in formula (4.8) the quantum theoretical equation[60] $v(n, n - \alpha) = \dfrac{1}{h} \{W(n) - W(n - \alpha)\}$. We thus obtain a three-component-chain theory: *wave theory – Maxwell's electrodynamics – quantum matrix mechanics*.

Let me now pose the following question: where does this three-component-chain theory connect up with nature? According to Einstein, such a chain ends up in observations in the sense of *production of sense impressions*. He declares (1969, 92, 94–95):

> Only theory decides what one can observe ... observation is in general a very complicated process. The process to be observed produces certain events in our measuring

---

58 For details of this see (Sommerfeld 1922, 157; 1923, 154).
59 What is also involved in the formulation of the equation for λ employed by Ångstrom is the knowledge embodied in the thermometer attached to spectrometer.
60 This equation goes back to the frequency condition introduced in (Bohr 1913b).

apparatus. As a result, further processes take place in the apparatus which finally ... produce a sense impression and fix the effects in our consciousness. Along this whole path—from the phenomenon to its fixation in our consciousness—we must know how nature functions, we must know the natural laws ... before we can claim to have observed anything at all. Only theory, that is knowledge of natural laws, enables us to deduce the underlying process from our sense-impression ... And we have to remember that the inference of representations and things from sense-impressions is one of the basic presuppositions of our thinking.

This, at least in my view, cannot hold for the following reasons. Into the equation for the magnitude λ, as shown already, enters magnitude b as distance, but distance as *magnitude* is not observable. Its meaning is given by the generalization of the *practical* procedure by which distances are measured, namely, by aligning one end of the measured object with one end of the object which serves as the measuring object, and, then, by finding out, if the other end of the measured object is also in the alignment with the other end of the measuring object, or if there is an overlap of one with respect to the other.[61] In addition, the *actual values* of λ, *and so neither of n nor of v* can be observed in the sense of perceiving. For example, the actual values of wave lengths of hydrogen's spectral lines are in the range of 10^{-7} mm. The same also holds for the actual values of magnitude b; they are not observable because of several thousands of lines engraved by means of the dividing machine on a piece of glass of the length of only approximately 20.3 mm. A practico-operational origin has also the meaning magnitude φ appearing in the equation for the wave length. The angle between two intersecting lines is given by the procedure of moving one of them about the point of their intersection until it coincides with other one.[62] So, *third*, that three-component-chain theory connects up with nature in the practico-experimental operations on nature.

Fourth, sensualism which was at work in the discussion on early quantum mechanics, reduced the epistemological concept of *practical-experimental operations with things* to that of observation in the sense of *passive sensual perception of things*. At the same time he did not take into account that here exists a type of measurement which is part of operations in mind employed in the process of theory construction and theory application to experiments. These epistemological

---

61 On this see (Ellis 1960). The measurement of temperature by Ångstrom by means of a thermometer goes back to the change of length of the column of the substance employed in the thermometer and thus is based also on the practice of measurement of distances.
62 Here I draw on (Feather 1968).

reductions can be seen in the reflections of Einstein stated earlier.[63] If my delineation of sensualism holds, then I can restate the following two statements of Einstein (which appeared in the discussion with Heisenberg). Instead of the statement that "Only theory decides what can be observed" (1969, 92) I state that *"Only an interchaining of theories decides what experiments we can perform at all"* and instead of "The inference of representations and things from sense-impressions belongs to the basic presuppositions of our thinking" (1969, 94–95) I state that *"The inference from experiments to representations and thinking about things belongs to the basic presuppositions of our thinking."*

Fifth, the clarification of the reductive nature of sensualism makes it possible to understand the nature of Heisenberg's *Anschaulichkeit* paper (1927), to explicate the epistemological aspects of the relations $p_1 q_1 \sim h$ and $E_1 t_1 \sim h$ given in this paper, and to deal with the views on this relation as presented in the discussion (Heisenberg et al. 1931).

In the beginning of the *Anschaulichkeit* paper, Heisenberg, in contradistinction to his *Umdeutung* paper, does not relate *Anschaulichkeit* to observability, but to *thinking* and *experiments as practical operations* on the things of nature under investigation. He declares that "We believe we understand a theory intuitively (*anschaulich*) when we can qualitatively think the experimental consequences in all simple cases and at the same time we have recognized that the application of the theory does not contain inner contradictions" (1927, 172). Since, in the beginning of the *Anschaulichkeit* paper he accepts the Kantian reading of "anschaulich," its translation should be "intuitive" and not "perceptible."[64] And, he does not view the actual values of the radiation magnitudes in the matrices as given in observation, but as "experimentally given numbers" (1927, 172).

Another interesting feature of the *Anschaulichkeit* paper is that Heisenberg grounds the respective magnitudes of quantum mechanics (for example p and q) in the practico-experimental operations on things and not in observations. And, in addition, he locates in this paper the uncertainty of the position and impulse of a microparticle in the *world*-, that is, *nature-changing character of human practico-experimental operations and not in the limitations of human perception, knowledge and cognition.*

So, in my view, the relations $p_1 q_1 \sim h$ and $E_1 t_1 \sim h$ are paradigmatic examples of a perfect, complete knowledge about the microobject given in the experimental

---

63 Such a reduction can be found also in the MIT-lectures of M. Born (1926, 69).
64 Here I deviate from the translation of "anschaulich" as "perceptual" given by Cassidy in (2009, 159).

operations of scientists. By subjecting a microobject to a radiation in order to obtain data about its physical parameters, physicists inescapably change its pre-experimental parameters. Whatever parameters the microobject had before it was experimentally influenced, the physicists cannot know them because they have as yet not produced experimentally any data on which we could then built our knowledge about this microobject. Its state before it was practically altered in an experiment is and will be unknown for them. *An entity which is not experimentally given to us is not knowable for us; it is a thing in itself.* My understanding of the term "thing in itself" differs from its Kantian understanding. *In order to start obtaining knowledge about a thing, the latter should not be passively perceived, if it can be perceived at all (and micro-objects surely cannot be), but has to be involved in active practical operations which are the point of departure of cognitive operations aiming at obtaining knowledge about the thing.*

Heisenberg, however, in his *Anschaulichkeit* paper often relapses into a sensualistic interpretation of quantum theory. After dealing with the impact of *experiments* on the physical characteristics of the microobject, he suddenly speak about *observation*, declaring: "I believe that one can succinctly formulate the emergence of the classical "orbit" as follows: The 'orbit' comes into being only through observation" (1925, 185) And at the very end of this paper he claims that the statistical character of quantum mechanics "is closely linked to the inexactness of all perceptions ... Physics ought to describe only the relations of perceptions" (1927, 197).

Such a relapse surfaces again in his statements in the discussion on quantum mechanics published in the journal *Erkenntnis*. Here one can read that while in classical physics the mechanical system (Heisenberg *et al.* 1931, 186–187)

> does not bother if it is observed. All observers find the same, namely, that the system's behavior is in accordance with the solution of the differential equation. Therefore one says that ... the perception is not important. In the quantum theory it is different. Here the observer has changed the system; the second observer thus sees it already differently. The system has no objective reality in the sense that each observer sees the same, so as each of them changes the system by his own observation. Therefore one cannot say that the wave function can be identified with the "thing in itself"; one has to say that it is a mathematical abstraction.

In my approach, contrary to the views presented by Heisenberg here and, partially, also in his Anschaulichkeit-paper, *the basis on which quantum mechanics is constituted are the practical operations – experiments – with the objects under investigation*, which are then *theoretically grasped by stating the matrices and the wave equation with the respective wave equation.*

Sixth, the statements of Heisenberg and Frank from *Erkenntnis* disclose yet another effect of sensualism in the epistemological discussion on the early quantum mechanics. Here I mean the *systematic denigration of theoretical cognition by denying any cognitive status to things in thought* produced and operated on in theory construction and theory application. Neither Heisenberg nor Frank take into account that the wave function can be viewed, following Schrödinger's explicit statement – as *thing of thought* (*Gedankending*) (1935, 811). Instead, Frank reduces the wave function to a "bond" between observables, while Heisenberg reduces it to a mere mathematical entity; so, both refuse to assign to it in (Heisenberg et al. 1931) a *physical referent*.[65]

In order to understand the possible referents of thought objects introduced in physics (e.g., wave function) I have to clarify certain post-Kantian themes as they appear in early quantum mechanics.

4.4 Clarifications: Some Post-Kantian Themes in Early Quantum Mechanics

In the *Umdeutung* paper Heisenberg viewed the concepts of space and time as referring to observables, and because in the "old" quantum theory of the inner-atomic realm they do not have such references (they are not applicable here), he created a form of quantum mechanics for this realm which was free of these concepts. From this he drew the conclusion that (1928/1984):[66]

> The Kantian a priori concepts form ... a strengthening and clarifying of those concepts which are adjusted to our daily experience. One should thus be inclined to believe that these Kantian concepts ultimately do not have their origin in the pure reason but in daily experience ... In the past, it was believed that one can strictly differentiate between the *content* of our thinking, of our experiences, and the *form* of our thinking, the "*a priori* content." That seems no longer true; rather, it is also the form of our thinking which is associated with the everyday experiences.

Schrödinger, on the hand, as shown above, holds the following complementary view: the concepts of space as forms of thinking (1926b, 509) are given *a priori* and cannot change; otherwise knowledge about the physical world would not be possible.

These opposing and complementary views on the concepts of space and time as forms of thinking (*Denkformen*) can be unified and reconciled as follows.

---

65 In my view, the wave function is not a *mathematical* construct. Even in its most simple form for one particle in a three dimensional position space it has the dimension [length]$^{-3/2}$, that is, it is a *physical* magnitude. On the dimension of the wave function see (Styer 1996).

66 For similar views on Kant's a priori concepts see (Heisenberg 1934).

Their unification I propose is as follows. These forms have an *innerworldly* status in the sense that humans acquire them by means of their practical operations in the world of nature unified with cognition/thinking including scientific cognition. But, once they are produced in thinking, they acquire an *outerworldly* status; they are applied to and drive both successive practical operations and thinking/cognition about nature, of course, up to the point when cognition at a certain stage yields mutually contradictory results which cannot be mutually reconciled any more. Thus, there exists a cyclic dialectics given in the concepts of space and time from the point of view of their innerworldly and outerworldly statuses: *they enable to operate practically in/on nature and think how it is and at the same time they are the result/product of practical operations in/on and of thinking about nature*. I consider this delineation as the *first revision* to which Kant's views should be subjected with respect to the epistemological discussions from the late 1920's on the new quantum mechanics.

This revision implies the following. The concepts of space and time in their *outer-worldly* aspect enable vision and perception in general as well as practical operations. Based on this I propose a *second revision* of Kant, namely, that space and time are the basic *categories* enabling the understanding of phenomena on which higher forms of understanding – based other categories (for example, causality) can be built. *They are categories of the thinking activities of sensual consciousness*.

From my epistemological interpretation of the $p_1 q_1 \sim h$ and $E_1 t_1 \sim h$ once combined with the above proposed first revision of Kant, a *third revision* – to be developed in the future – can be proposed. While Kant provided a philosophical reconstruction of the categories of understanding as entities enabling to think an object, an epistemological reconstruction should yield also a *reconstruction of the structure of practico-experimental operations on and with objects of nature*. Only by unifying these two reconstructions one can obtain a coherent epistemological understanding of the interweaving of practico-experimental operations and of operations in mind as given in physics.

This unification, in turn, requires a modification of Kant's understanding of the term "experience" (*Erfahrung*). It should not be restricted to cognition and thinking related only to perceptually given objects but should be understood as covering both *practico-experimental operations on things in the world of nature* as well as *thinking performed on the things in thought referring to the things given in nature*.

The delineation of this third revision enables one to understand one of the sources of the views symptomatic of sensualism, namely, that the whole scientific enterprise is reduced to the dichotomy of *theoretical* and *perceptual*, and where the entities dealt with in theories are viewed as mere mathematical entities devoid of any objective content. The source of this is located in the lack of an epistemological reconstruction of practico-experimental operations in science going hand in hand with the epistemological reduction of these operations to the Kantian concept of experience (*Erfahrung*) in the sense of passive perception, as well as the neglect of a detailed epistemological investigation into theoretical procedures employed in physics.

The absence of an epistemological differentiation between the concept of practico-experimental operations and the (Kantian) concept of experience (*Erfahrung*) becomes apparent when one compares the paper (Born – Heisenberg 1928b), presented in French by Born and Heisenberg at the Solvay conference, with its original typescript written in German (Born – Heisenberg 1928a).[67] In the typescript the authors state "Von einer in diesem Sinne anschaulichen Theorie kann aber nur verlangt werden, dass sie ... für alle in ihrem Gebiete denkbaren Experimente die Ergebnisse eindeutig vorherzusagen gestatte" (1928a, 1). In the French translation, however, the German "Experimente" is replaced by the French "expériences," that is, by the French synonym of Kant's term "Erfahrung," and the whole statement then is "D'une théorie intuitive en ce sens on doit donc demander ... qu'elle permette de prédire sans ambiguïté les résultats de toutes les expériences imaginables dans son domaine" (1928b, 144).

The other source of the views symptomatic of sensualism can be understood in epistemological terms as follows. Let me take one term from the pre-quantum formula (4.8), that is, Balmer's formula in its Rydberg-Ritz restatement multiplied with the constant c, as well the formula (4.3) derived by Schrödinger and divide the latter by h and substitute $h/2\pi$ for the constant K. We then obtain the following two formulas:

$$-v = c \cdot N_0 \frac{1}{m^2}, \quad -v = \frac{2\pi^2 m e^4}{h^3} \frac{1}{l^2} \tag{4.9}$$

Comparing them reveals that the claim of P. Frank, namely, that the "bond between the observations is established by means of a nonobservable mathematical scheme, namely, the wave function." (Heisenberg *et al.* 1931, 185) does not hold. Neither the magnitude v in the left equation in (4.9), as shown above, nor the v in the right equation derived from the equation (4.2) for the ψ-function refer to something observable. Likewise, neither the magnitude e (charge of the electron)

---

67 For the English translation of the German typescript see (Born and Heisenberg 2009).

nor Planck's constant h entering into the formula for v refer to something observable. Their actual values are not observable either; this was known already in the mid 1920's (Birtwistle 1926, 25), to be in the range of 10^{-10} (in units E.S.U.) and 10^{-27} (in units ergs.sec.), respectively.

So, from the point of view of *unobservability*, there is no epistemic difference between the ψ-function and both formulas in (4.9). But there exists an important difference in the epistemic/cognitive status of this function with respect to the status of the formulas in (4.9). The ψ-function refers to the quantum-mechanical state of the system under investigation, where the change of the state (for example, by the eigendynamics of the system and/or by the impact of another quantum-mechanical system) is expressed by subjecting the ψ-function to a transformation in the wave equation. This change stands for the *ground* of the emitted spectral lines. So, the ψ-function with its equation expresses the knowledge about the ground of spectral lines, where the latter are expressed by both formulas in (4.9).

I thus arrive at the *fourth revision* of Kant. As long as the epistemological term "thing in itself" is not understood as a thing outside cognition, but as a type of thing in thought, namely, as the *thought ground of phenomena*, it can be known by means of thinking and can be subjected to transformations in mind.

This requires performing a *fifth revision* of Kant's theory. While he provided in the *Transcendental Analytics* an epistemological reconstruction (in the form of a table) of categories enabling to think things as *appearances*, we are now in need of an *epistemological reconstruction of basic concepts enabling to think and cognize a thing as the ground of appearances*. So, as claimed above, if the concepts of understanding (categories) stand for the thinking activities of the sensual consciousness and for the thinking about the appearances of things, then the subject matter of transcendental dialectics should not be God, mind and world (as a whole), but the concepts of reason, I label them as "categories," enabling to think and thus know the very ground of appearances.[68]

This change of the framework and enlargement of the content of Kant's epistemology, (i.e., the renouncement of the claim that we cannot know, even if we can think, the ground of appearances) leads to *seventh revision*. If the various spectral lines are the appearances of the quantum states of the system producing these lines, then we have to distinguish – given the two forms of Balmer's formula in (4.9) – between *two forms, levels of the knowledge of appearances of*

---

68 Here I deviate from Kant also terminologically; the concepts of reason are labeled by him as "ideas."

the ground: one which *predates* the knowledge of the ground and one which is *derived from* the knowledge of the ground.

Due to his claim that we cannot know the ground of phenomena, Kant dealt only with the constitution of empirical intuitions (by the employment of space and time as pure forms of intuition on perceptions) and with the movement of cognition from empirical intuitions to appearances (by the employment of categories), while the movement from appearances to their ground dropped out of his reconstruction. However, once we claim that in cognition one moves, by means of thinking, *to* and *from* the knowledge of the ground, we are in need of a epistemological distinction between two modes of knowledge and at the same time cognition of phenomena.

If we stick to the term "appearance," as referring to the knowledge of phenomena which precedes the knowledge of the ground of these phenomena, then we are in need of yet another term which would refer to the knowledge of phenomena once they are derived from the knowledge of their ground. The movement of cognition to and from the ground of the phenomena was considered by C. G. Hempel, who declared (1970, 142):

> Theories are normally constructed only when prior research in a given field has yielded a body of knowledge that includes empirical generalizations or putative laws concerning phenomena under study. A theory then aims at providing a deeper understanding by construing those phenomena as manifestations of certain underlying processes governed by laws which account for uniformities previously studied.

As can be readily seen here, he assigns to the phenomena once they are derived from some underlying processes, the latter being labeled by me as "ground," the term "manifestation." I thus propose to differentiate between *appearances* and *manifestations* as *two forms of knowledge of phenomena*.

The *epistemological* differentiation between these forms of knowledge of phenomena becomes understandable once we compare the two formulas in (4.9). The quantum form of the Balmer formula differs from its prequantum form by the presence of Planck's constant h. And where this constant comes from? The answer in the terms of philosophy of science is: the equation (4.2) is part of the *explanans* from which the quantum form of the Balmer formula was derived as the *explanandum* and the constant h was shifted from the former to the latter. The pre-quantum form of the Balmer formula does not contain the constant h because it was derived in the context of prequantum wave theory and prequantum (classical) electrodynamics.

Based on my distinction *appearance – ground – manifestation* I can provide an overall and comparative epistemological characterization of matrix and wave mechanics. What unifies them, from the point of view of epistemology, is that they *do not pertain to observables but to nonperceptually given things in thought*; both have the status of a *theory*. The difference between them lies in the fact that matrix mechanics is, when *compared* with wave mechanics, more closely related to spectral analysis: to the experiments producing spectra and to its description of spectra by means of terms like "frequency," "wave length," "combination principle." From this point of view, matrix mechanics and wave mechanics can be classified as standing for *phenomenological* and *post-phenomenological* knowledge, respectively.

My understanding of the term "experience," as well as of the relations among experiments, spectral analysis, matrix mechanics and wave mechanics I express by means of the following figure:

Figure 4.4: Relations among experience, experiments, spectral analysis and quantum (matrix and wave) mechanics

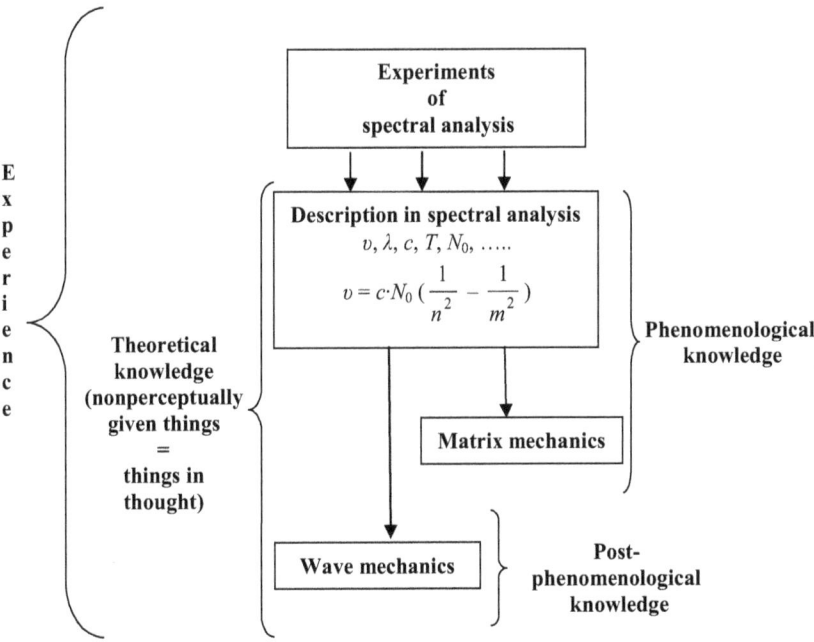

I claimed that the ψ-function as a thing in thought is assigned to the state of the system under investigation. Here one could, seemingly, claim in a Machian manner that this function is still an empty construct, a "mere thing in thought" because it lacks any reference in the realm of observables appearances. As shown above, the systems under investigations for which the respective wave equation in

123

the first through fourth *Mitteilung* should hold are themselves idealized entities, that is, *things in thought*.

My answer to this Machian objection is as follows. Not only, as shown, does the knowledge of appearances (e.g., Balmer's formula in the pre-quantum form) refer to a nonobservable, but in the movement of cognition and thinking characterized by the trio of categories *appearance – ground – manifestation*, the introduction of the ψ-function with its wave equations is accomplished at a level of cognition where the phenomena as appearances temporarily drop out of cognition. By introducing the ψ-function with its wave equations, Schrödinger abstracted in mind from the phenomena – the line spectra of atoms – that is, from *appearances*, and turned to the ground of them, to the quantum mechanical state of *the atomic system* producing the respective energy spectra. Stated otherwise, *the wave equation is non-vacuously fulfilled by the ground of the respective physical system to which we have in mind reduced this system.*[69] Only then did Schrödinger return to the phenomenal behavior of the atoms by deriving the respective eigenvalues of the system from the respective wave equation, where these eigenvalues already have, in the context of this derivation, the epistemic status of *manifestations*. So, what is *understood as real* does change in the advancement of cognition from appearance via ground to manifestation.

Accordingly, one of the main deficits of sensualism is that it views thinking about (or, to be more precise, *perception* of) appearances as the only possible way to know the real. The standard objection of sensualism against the attempt to move via appearances to their ground, namely (i.e., that it is *based on the misleading claim that while the knowledge of the ground yields a true understanding of the real, the knowledge of the appearances yields "untrue" forms of knowledge*) does not hold with respect to my views presented here. In my approach, I view knowledge about appearances as the necessary and only possible point of departure in order to obtain knowledge about both their ground and about the mechanism by means of which the phenomena of the ground are produced. Kant was, from this point of view, right when he emphasized that the knowledge of appearances does not have the status of an illusion (A293=B349–350; 1998, 384); this knowledge has an *objective content*. If it did not, it could not serve as the point of departure for the thought movement to the ground and its manifestations.

---

69 The most simple thought object for which a wave equation is non-vacuously fulfilled is a system reduced in mind to a free microobject not subjected to an external force field. Such a thought object was not explicitly construed by Schrödinger in his *Mitteilungen*. One obtains the wave equation for this thought object by putting the expression e^2/r in the equation (4.2) equal to zero.

Chapter 5: Measurement and Conceptual Networks in Early Thermodynamics

The aim of this chapter is to provide a philosophical-cum-methodological analysis and reconstruction of the conceptual development in early thermodynamics spanning the period from the late 17th century to the mid-19th century.

I start with an analysis of the understanding of the terms *fire/heat* and *temperature* in the works of E. Halley, B. Taylor, J. T. Desagulier, E. Boerhaave, J. Black, J.-A. De Luc, J. C. Wilcke, Lavoisier and Laplace, and show how they approach the relations between the magnitudes expressed by these terms and how they deal with the issue of measurement of temperature and heat. Next, I reconstruct the bootstrapping nature of measurement of heat/fire based on the measurement of temperature and show how Joule and Thompson escape this bootstrapping by conceptually treating the production of heat itself.

Then, I provide a philosophical reconstruction of thermodynamics of this period by means of the category pairs *cause-effect, reason-reasoned* and *mesurans-mesurandum*. Finally, I shall test the efficiency of the employment of these category pairs with regard to thermodynamics by showing that in this way one can methodologically evaluate W. Thomson's introduction of the concept of absolute temperature.

In order to prevent any possible misunderstanding, I would like to emphasize that here I do not deal with the disputes about the nature of heat which took place in the 18th and 19th centuries.

5.1 Fire, Heat, Temperature, Thermometer and Weight × Distance

5.1.1 Fire/heat and temperature

Not later than by the end of the 17th century, the terms playing a central role in the gradual formation of thermodynamics were *heat* and *cold* viewed as qualities which can be subjected to quantification by measuring their effects on the quantitative characteristics of the bodies on which they act. So, according to E. Halley (1693, 650):

> *Qualities*, such as heat and cold ... are not otherwise to be estimated, but by their effect on the *quantity* of some body they act on, increasing or lessening the dimensions of thereof; or else by the motions they produce, both which subject them to mensuration.

125

The chief reason for the employment of such an *indirect* method of quantification was the fact "that we know not the causes of the expansion of fluids by heat, or of their contraction by cold" (1693, 650). At the same time, this indirect character of quantification of heat and cold was viewed as problematic because, due to the nature of liquids employed in the instruments measuring that effect (1693, 650–651, 655):

> the same degree of heat does not proportionally expand all *fluids*, some swelling with a gentle warmth, and others not till they be considerably hot; some boiling with a moderate heat, and others not at all. Some capable of great expansion, others increasing very little; so that it may well be concluded that none of them does increase or diminish in the same proportion with heat [...] this power of dilating or contracting with heat and cold is as specifically in them as their gravity, refraction, etc.

Even with this realization of the lack of knowledge about the nature of heat and of the problematic nature of the measurement of heat (by means of the measurement of the expansion/contraction of fluids), still the view shared by those who constructed and employed instruments for such a measurement was that these instruments were indicating the quantity of heat. So, for example, B. Taylor declared that "[i]t has ... been gradually supposed ... that the expansion of the liquor in the thermometer, is proportional to the increase of heat" (1723, 291). Therefore, the instrument – contrary to the fact that it was perfectly clear that it indicates the amount of heat only in an *indirect* manner and that it indicates in a *direct* manner something different from heat – was nevertheless labeled as "*thermo*meter" understood as referring to "an instrument contrived for measuring the degrees of heat in bodies" (Desaguliers 1744, 289).

The subsequent development of early thermodynamics went along two different, but closely interrelated lines. One focused on the development and elaboration of the methods of construction and conceptual grounding of the functioning of precise thermometers as well as for reflections on the possible relations (e.g., proportionality) of the quantity directly indicated by the thermometer to the cause of this indication – heat. The second, based on results of the first line of development and the practical employment of thermometers, tried to show that a thermometer really points to something which is different from that what is directly provided by the reading on the thermometer itself.

The first line of development is readily seen in R. Boerhaave's *A New Method of Chemistry*. Here, in the section *Of Fire*, the meaning of the term "fire" is being employed in the sense of one of the four (ancient world-view) elements and where fire is viewed as different but still related to heat, so that it holds that (1743, §12, 210–211):

> Heat ... is in the first place attributed to fire, and with good reason, as the two are connected by a very close tie; yet when we look near into the idea of heat, it will easily appear that by this word men denote a certain sensation impressed in the mind, when their organs of feeling happen to undergo some change by the application of fire to them; but in this idea there is nothing contained that denotes either an action of fire, or the change of the feeling organ in the body ... I learn nothing by that concerning fire.

It appears, then, that what Boerhaave is looking for instead is a more reliable sign/mark for fire; one that is inseparable from fire, peculiar to it and also obvious to the senses of all persons. As the chief sign of this he identifies the "expansion or rarefication both of fluid bodies and solid" (1743, §11, 210), so that (1743, §16, 214):

> this expansion of bodies by heat, if it be performed in a glass hermetically sealed, cannot arise from any other physical cause hitherto known, but from fire alone, so that the thing desired is now found, viz. a true, certain, inseparable, peculiar, characteristic of fire; and this alone we shall use in the sequel for discovering the nature of fire; taking it for granted, that in all phenomena where this rarefication is observed, there fire is discovering itself to us in proportion: by which we shall have an opportunity of examining fire in almost all its conditions, and of reasoning about its hidden nature, which discovers itself in all the experiments of this kind.

So, it becomes apparent that the concept of fire is in Boerhaave's approach Janus-faced: it is still the historical antecedent of the concept of heat, where the latter developed from the former, but at the same time the latter comes very close resembling the former, since both display the same effect – a change of the shape of fluid and solid bodies; accordingly, they cannot be strictly separated any more. He, therefore, declares that *"All bodies expand with heat"* (1743, §22, 216) and *"Expansion* [is] *the measure of heat"* (1743, §25, 216).

The same line of development can also be discerned in the section *"Du Thermomètre"* of J.-A. De Luc's *Recherches*. He states (1784, §410a, 55–56):

> The increase of volume of a body due to the increase of its heat is the overall, and perhaps unique, means to measure the effects of this cause: and fluids are always to be preferred to solid; both because the heat produces in them the greatest effects and because they can make the increase of their volume more perceivable by forcing them to expand in narrow channels,

Under the term "heat" he understands "the fire, or a cause, one of the effects of which is the expansion of bodies" and at the same time he expresses "by the word *temperature* the state of bodies only with respect to *heat* alone" (1784, §410b, 56, note (a)).

De Luc explicates also the principles upon which a thermometer should be constructed, declaring (1784, §411, 57–58):

If we could provide a fluid which measures equal changes of heat by equal variations in its volume, this fluid ought to be preferred to all the other in thermometers, because it would indicate to us incremental amounts of increase or decrease of heat; and consequently the real relationships between these quantities.

Based on experiments he then finds out that "mercury is from all the liquids employed in thermometers that one which measures most exactly the differences of heat by the differences in its volume" (1784, §421z, 150).

The second line of the development of early thermodynamics stood for the process of gradual differentiation between the concept of heat and the concept of temperature. In this process a crucial role was played by the works of J. Black, G. W. Richmann and J.-C. Wilcke. Here I have in mind Black's differentiation between *quantity of heat* and *intensity of heat* (i.e., that is, what is now labeled as "temperature"). He states that (1807, 75):

> these are two different things, and should always be distinguished, when we are thinking of distribution of heat. If, for example, we have one pound of water in one vessel, and two pounds of water in another, and these two quantities are equally hot, as examined by the thermometer, it is evident that the two pounds must contain twice the quantity of heat that is contained in one pound.

This difference was approached in the thermodynamics of the 18th century by means of the concept of *latent heat* by Black in his lectures and by Wilcke in (1774), who drew upon on a formula stated by Richmann in his paper "*De quantitate caloris*."[70]

In the latter, Richmann dealt with the mixing of two fluids, where to these fluids and to the result of their mixing, he assigned the terms "*calor*," "*temperies*," "*caloris gradus*" and "*calor generates*" in a highly undifferentiated manner indicating thus that he did not as yet differentiate between heat and temperature. The point of departure for stating a formula describing that mixing is as follows (1747–1748, §1, 152–153):[71]

> I imagined, that the *calor* in a fluid of a certain *temperies* is equally distributed in the whole mass of this fluid, and at the same time thought that if the same *caloris gradus* is distributed in masses 2-times, 3-times, 4-times, etc. than this mass, then the *calor hinc generatus* has to be 2-times, 3-times, 4-times, etc. smaller than the initial [*calor generatus*], and in general, the same *calor* is inversely proportional to the masses in which it is distributed.

---

70 For Black's introduction of the term "latent heat," see (Black, 1807, 122); for a detailed analysis of the introduction of this concept into thermodynamics see (McKie and Heathcote, 1975).

71 In order not to impute to Richmann the differentiation between heat and temperature, I do not translate his Latin terminology.

Based on this he derived a formula stating the following (1747–1748, §2, 153):

In this way, *calor* = *m* of mass *a*, and *calor* = *n* of mass *b* are equally distributed in mass *a* + *b*, and *calor* in this mass, in the mixture of *a* and *b*, has to be equal to the sum of the *calor m* + *n* distributed in the mass *a* + *b*, or = $\frac{ma+nb}{a+b}$.

This formula was initially accepted by Wilcke, who restated it into the form $\frac{mc+MC}{m+M}$ and which yields a quantity he labeled as "thermometer-degrees" (*Thermometergraden*) *tg* (1772a, 98; 1772b, 93).

Initially, Wilcke – like Richmann – did not differentiate in any strict sense between the concept of heat and the concept of temperature, as can be seen from the meaning he assigned to the signs "*c*" and "*C*", namely, degree of heat, but to which he, then, assigned values obtained from thermometer readings. And he viewed as the goal of his endeavor to discover a law or rule for mixing water with snow "according to which one could know in advance for all cases the *heat of the mixture*, or the *thermometer-degrees*" (1772a, 99; 1772b, 94 – emphasis is mine).

Later, however, he began to separate the concept of heat from the concept of temperature, because he found out that while it holds for the mixing of warm water of mass *m* with a temperature *c* with water of mass *M* with *C* = 0° degrees of Celsius, it does not hold once snow, instead of water, with mass *M* at 0° degrees of Celsius is mixed with warm water of mass *m* with temperature *c*.

Based on this realization, he performed a series of experiments in which he mixed snow of a certain mass at 0° degrees Celsius with a mass of water at a certain temperature, so that the ratio of their masses was initially 1:1, then 1:2, ..., and finally 1:16. Then, he measured the temperature of the resulting mixture. Based on the formula for *tg* he also computed what would have been the resulting temperature of mixture composed of water at 0° degrees Celsius with water at the same temperature as the water in the experiment with snow, and again in mutual ratios 1:1, 1:2, ..., 1:16 of their masses.

Wilcke viewed the difference between the actually measured temperatures of the water-snow mixture and the computed temperature *tg* of the water-water mixture as a measure for the heat lost by the snow at 0° degrees when it turned to water with the same temperature. By statistically treating the results of his computations, he arrived at the average value of 72 degrees for that difference. This result he summarized as follows (1772a, 105; 1772b, 101):

> *in a mixture of water and snow, heat is not divided without loss and does not stay active and residual, as it is in the case of a mixture of water and water, but instead a certain and constant amount, corresponding to 72 degree on the thermometer, is always lost. From this point of view, one can regard the snow as being in a negative state; it attracts heat*

or destroys it, without being warmer from it, but instead becomes liquid. *Thus, these 72 degrees are spent only on liquefaction, after that the snow behaves as ice-cold water, and enables a uniform distribution of the remaining heat through the whole mass.* So, if a certain amount of water has 72 degrees of heat and an equal amount of snow is added, then all this heat is taken and the mixture becomes ice-cold. If the water's heat is 98 degrees, then 72 degrees are lost and the remaining 26 degrees are divided among two equal masses, so only 13 degrees are left in the mixture.

The general conclusion he then draws in distinction to Richmann's conclusion, is that heat and temperature are different entities, where the latter "is a *real matter, whose amount can be measured,* whose lack or excess transforms a body from solid into liquid, which can be often found in the bodies without being detected by distribution on a thermometer, but which can be again separated from them" (1781a, §3, 50–51; 1781b, 50–51).

So, the development of the early thermodynamics as presented above yielded the following results. First, it led gradually to a clear separation of the qualities of temperature and heat (fire). Second, it led to the realization that their relation is that of an *effect* to its *cause*. Third, it yielded the realization that the cognition of the latter is performed on the basis of the former in the sense that knowledge about the quantity of the former yields knowledge about the latter. Fourth, this way of acquiring knowledge about the quantity of heat came to be viewed by scientists, starting from the late 17th century throughout the 18th century as the *only* way to measure it because of a lack of knowledge about the cause of heat itself. So, for example, Lavoisier and Laplace in their common *oeuvre* declare the following (1784, 288):

> In the ignorance in which we find ourselves as to the nature of heat, we are left only with the option to well observe its effects, where the most important are the extension of bodies, the liquefaction and the conversion into vapor. Among these effects the easiest to measure should be chosen and which is proportional to its cause; this effect would represent the heat.

Black characterizes in a similar manner the situation when two bodies unequally heated are approached to one another, as a situation when (1807, 28)

> the warmer or less cold body ... acts on the other, and communicates to it a real something which we call heat ... it may perhaps be required ... to express more distinctly this something, to give a full description or definition of what I mean by the word, heat in matter. This, however, is a demand which I cannot satisfy entirely ... our knowledge of heat is not brought to the state of perfection that might enable us to propose with confidence a theory of heat, or assign an immediate cause of it.

Based on these four results of the development of thermodynamics, let me know reconstruct the bootstrapping nature of measurements involved here.

5.1.2 Bootstrapping in the measurement of heat by means of the measurement of temperature

As shown above, the search for a suitable substance which could be employed in a thermometer was driven by the requirement that it should fulfill the condition that "it measures the same variations of heat by the same variations of its volume" (De Luc 1784, §411, 57).[72]

The idea of the *same* amount of heat being involved both in experiments and practice of measurement played an important role in the thermodynamics of the 18th and early 19th centuries. So, for example, Boerhaave declared that (1743, §17, §22, §23, 214, 216):

> Fire expands even the hardest bodies, in all their dimensions, so long as it is contained in them ... It must be observed that this expansion of solid bodies by fire is general, and obtains after this manner in all its instances where observations has been made ... But you must not imagine, that it is true in all bodies. On the contrary, in the heaviest bodies it appears the least, supposing the fire the same; and in the rarer the greater.

From this he, then, derived the general rule that "all the expansions of bodies, by the same fire, be to each other as their [specific] weights" (1743, §23, 216).

The issue of the relation between the behaviors of thermometric substances to the same amount of heat was investigated also by Gay-Lussac. He stated that (1802a, 138; 1802b, 208):

> The thermometer, as it is given today, cannot serve for the indication of exact proportions of heat, because we do not know as yet what relation exists between the degrees of the thermometer and the quantities of heat which they should indicate. It is, indeed, generally thought that the equal divisions of its scale represent the equal tensions of caloric; but this opinion is not based on any really positive fact.

Then, drawing on a series of experiments which were described in detail (1802a; 1802b) he drew the following conclusion: "All the gases (whatever maybe their density and the quantity of water they hold in solution) and all vapors equally dilate by the same degree of heat" (1802a, 174–175; 1802b, 267).

The question one faced in thermodynamics of the 18th and early 19th century was *how to determine that a certain kind of thermometric substance dilates (contracts) in the same proportion as is the proportion in which this substance absorbs (looses) the same amount of heat, once, at that time, the only way to find out the latter proportion was by employing a thermometer based on the use of substance?*

---

72 For a similar statement of this requirement see (Dulong, 1817) and (Dulong and Petit, 1816).

This is the bootstrapping feature of the employment of the thermometer readings for the measurement of heat being at work.

Even if in the 18th and early 19th centuries there seemed to be no other way to find out the quantity of heat at work by any means other than a thermometer, one was in fact employed even if initially not subjected to conceptual reflection. Here I mean the *practical production* of heat used in experiments. Strangely enough, in the description and schematic representation of experiments performed, for example, by Gay-Lussac (1802a; 1802b), the warming of water which was then used to warm the substances whose dilatation should be measured, was completely left out of descriptions and schematic representations. Such a description could have involved, for example, a detailed description of the weight amount of a particular kind of wood of a certain age or of a specific kind of coal burned under certain conditions in a concrete experimental set up.

This production, once conceptualized – as I will now show – made it possible to solve two problems at once: to conceptually identify the origin of heat *and* to escape the above described bootstrapping in the measurement of heat by quantifying heat independently from both the employment of a thermometer and the knowledge about the relation of temperature and heat. This conceptualization and quantification was based on the use of certain concepts of mechanics.

5.1.3 Work as weight × distance

A. T. Petit in (1818) considered the case when a machine, due to the action of a force, lifts a body to a certain height. The effect of this force in such a process "depends both on the weight shifted and the height to which it is shifted" (1818, 289). The expression he then employs for such an effect is gMH, where "g" stands for the intensity of gravity in the location where the machine is placed, "M" for the mass of the body lifted and "H" for the height to which it is lifted. In addition, he viewed this effect as transformable into the *vis viva*, understood by him as equal to $\frac{1}{2}MV^2$, of a body with mass M acquiring a velocity V when falling from a height H, so that holds $gMH = \frac{1}{2}MV^2$.[73]

The conceptualization of the lifting of a body by force expended by a machine was generalized by G. G. Coriolis to the case when the body is moved along a certain path and, based on this, he introduced the term "work." Accordingly, he declared the following "I designate by the word *work* the quantity which is quite commonly called *mechanical power, quantity of action* or *dynamic effect*"

---

73 For a more detailed analysis of the introduction of the concepts of power and work see (Cardwell 1967).

(1829, iii) and: "One attaches to the word 'work' the idea of an exercised effort and of the path covered simultaneously ... This name is uniquely suited to designate the unification of two elements: path and force" (1829, 17). In addition, he assigned to the word "*vis viva*" a new meaning, namely, that of $½MV^2$ (1829, iii).

Based on these conceptual innovations, the experiments performed by J. P. Joule, compared to experiments performed earlier, for example, by Gay-Lussac, underwent a profound change. The practical processes of heat production involved in them were now driven by the concepts of mechanical power (work) and *vis viva* brought in from mechanics and, they were already theoretically reflected upon by means of these concepts. Therefore, Joule could also reflect both conceptually and metrically ("quantificationally") on the chain *production of heat – change of heat – change of temperature*.

5.1.4 Joule's on the Mechanical Equivalent of Heat

Already in the paper (1843), Joule presented the idea of the production of heat as follows: "when we consider heat .... as a state of vibration, there appears to be no reason why it should not be induced by an action of a simply mechanical character, such, for instance, as is presented in the revolution of a coil circle before the poles of a permanent magnet" (1843, 263). And the general conclusion he arrives at as to the origin of heat was that "wherever mechanical force is expended, an exact equivalent of heat is always obtained" (1843, 43).

Then, in the paper (1847), he related heat to living force and force of attraction in the sense of force of gravity claiming that these (1847, 271)

> are mutually convertible into one another. In these conversions nothing is ever lost. The same quantity of heat will always be converted into the same quantity of living force. We can therefore express the equivalence in definite language applicable at all times under all circumstances. Thus the attraction of 817lb. through the space of one foot is equivalent to, and convertible into the living force possessed by a body of the same weight of 817 lb. when moving with the velocity of eight feet per second, and this living force is again convertible into the quantity of heat which can increase the temperature of on pound of water by one degree Fahrenheit.

This, in his view, expressed the "knowledge of the equivalence of heat to mechanical power" (1847, 271), that was, in fact what Coriolis, as shown above, labeled as "work."

From this he drew the following conclusion as to the very nature of heat (1847, 273–274):

> We have ... shown that heat can be converted into living force and into attraction through space ... Heat must therefore consist of either living force or of attraction through space.

In the former case we can conceive the constituent particles of heated bodies to be, either in whole or part, in a state of motion. In the latter we may suppose the particles to be removed by the process of heating, so as to exert attraction through greater space. I am inclined to believe that both of these hypotheses will be found to hold good – that in some instances, particularly in sensible heat, or such as indicated by the thermometer, heat will be found to consist in the living force of the particles in which it is induced.

Finally, based on experiments in which he produced heat in liquids (water, mercury) and solids (cast iron) by subjecting them to friction by means of a wheel driven by the descend and ascend of interconnected weights along a certain path, he arrived in (1850) at the following two conclusions (1850, 82):[74]

> 1st. *That the quantity of heat produced by the friction of bodies, whether solid or liquid is always proportional to the quantity of force expended*. And,
>
> 2nd. *That the quantity of heat capable of increasing the temperature of a pound of water (weighed in vacuo, and taken at between 55 and 60) by 1 Fahr[enheit], requires for its evolution the expenditure of a mechanical force represented by the fall of 722lbs. through the space of one foot.*

These two conclusions enabled him to find answers to two questions which were stated by the previous development of thermodynamics. 1) *What is the cause of heat?* was answered by Joule as follows. *The cause of heat in a body is the living force of its particles.* 2) *When one can speak about the same change of heat?* was answered as follows. *Same change of heat stands for same change of living force.* This second answer thus provided also an escape from the above reconstructed case of bootstrapping.

5.2 A Philosophical Explication

Let me now make an attempt at philosophical explication of the above reconstructed pre-Joulian thermodynamics as well as of Joule's conceptual and quantificational chain *production of heat – change of heat – change of temperature*. As an instrument of such explications I shall employ the following three pairs of categories: *cause-effect, reason-reasoned* and *mesurans-mesurandum*.[75] I employ these pairs to approach from the point of view philosophy of science the following three issues: a) the *direction of causation* as understood by thermodynamics in the 18th and 19th centuries, b) the *direction of reasoning* performed

---

74 For a reworking of these experiments see (Sibum 1995).
75 On the status of categories see (Walsh 1953) and (Hanzel and Černík and Viceník 1994). On the pair *mesurans – mesurandum* see (Pawson, 1989).

in this thermodynamics, and c) the *direction of the operations of measurement* employed in this thermodynamics.

The categorial basis of the pre-Joulian thermodynamics as it was formed in the 18th century and in the first decades of 19th century can be explicated in philosophical terms as follows. In this period, on the one hand, it was presupposed that there exists a causal nexus in the sense that a change of heat as a *cause* brings about a change of temperature as its *effect*; that is, it was claimed that the change of heat as the cause is the basis on which is based its effect – the change of temperature. This direction of understanding of the causal determination I symbolically represent as follows: $\Delta Q \xrightarrow{c} \Delta t$; where "$\Delta Q$" stands for change of heat, "Δt" for change of temperature and "$\xrightarrow{c}$" for supposition about the existence of causal determination.

But, on the other hand, in the *reasoning* of that thermodynamics, the change of heat as the presupposed cause was the endpoint of this reasoning – i.e., the *reasoned* for short, while the change of temperature as the presupposed effect was the starting point of this reasoning – i.e., the *reason*. So, in order to find out that a change of heat takes place, pre-Joulian thermodynamics gave as the reason *that* that should have been caused by heat, namely, the change of the indication on a thermometer, i.e., a change of temperature. And, because in that period of the development of thermodynamics the indication of change of temperature was viewed as the sole way to get access to the change of heat, the understanding of the change of heat and the understanding change of temperature did not stand in fact for two mutually different contents (forms, ways) of knowledge. The direction of reasoning involved here I express symbolically as follows: $\Delta Q \xleftarrow{r} \Delta t$, where "$\xleftarrow{r}$" indicates the direction of reasoning from the reason to the reasoned.

These features of the understanding of the relation between change of heat and change of temperature also manifest themselves also in the approach to the measurement of changes of the magnitudes heat and temperature. Here we have a situation when, on the one hand, a quantity assigned to a change of temperature serves as a measuring entity – i.e., *mesurans* – for the quantity assigned to a change of heat which has, with respect to that quantity, the status of a measured entity – i.e., *mesurand*. This direction of measurement involved here I express as follows: $\Delta Q \xleftarrow{m} \Delta t$.

In the relation between change of heat and change of temperature we find relations which are going, with respect to each other, in opposite directions. Change of heat as the cause should be *that* according to which one should understand the change of temperature as its effect, but, at the same time, the change of heat is both the *reasoned* and the *mesurandum*, i.e., *that* about which one reasons about

135

and measures by means of change of temperature as the *reason* and the *mesurans*. So, heat – instead of being understood in theory as something causally prior with respect to change of temperature and, thus, as independent from the latter and which it causes – is reasoned and measured by the change of temperature and in fact derived from it.

Stated otherwise: in the pre-Joulian thermodynamics the *supposition about the existence of causal determination and about its quantitative characteristic (about relation of the quantity of the cause to the quantity of what it causes) went into the opposite direction as compared to the direction of reasoning and measurement by means of which knowledge about the cause and its quantity was acquired, namely, from the knowledge about its effect and the quantity of this effect*. This bi-directionality as given in the pre-Julian thermodynamics I express symbolically as follows ("$\xleftarrow[r\&m]{}$" stands for the direction of reasoning unified with the direction of the operation of measurement):

Figure 5.1: *Supposition about the direction of the causal determination of the change of temperature and the directions of reasoning and measurement in pre-Joulian thermodynamics*

$$\Delta Q \underset{r\,\&\,m}{\overset{c}{\rightleftarrows}} \Delta t$$

Let us now turn to Joule's conceptual and quantificational chain *production of heat – change of heat – change of temperature*. If we examine, from the point of view of the two answers given at the end of section 3, the 1st and 2nd conclusions on Joule's part stated above then the following issues are worth explicating by means of categories.

First, Joule in the 1st conclusion views force expended on a body as the cause of the change of heat. In addition, in the 2nd conclusion the measurement of ΔQ draws on the measurement of force expended on a certain path, that is, of work W. This I express as $W \xrightarrow{r\,\&\,c\,\&\,m} \Delta Q$; here "$\xrightarrow{r\,\&\,c\,\&\,m}$" stands for *direction of reasoning about causation* unified with *supposition about the existence of causal determination* and *direction of measurement*.

Second, he acknowledges in the 2nd conclusion that there exists a causal nexus between the change of heat and the change of temperature in the sense that a change of the former causes a change of the latter. In addition, Joule accepts from the previous ("pre-Joulian" phase of) development of thermodynamics the fact that the quantity of heat is measured by the quantity of temperature, that is, the former is the *mesurandum*, while the latter is – with respect to the former – its *mesurans*. This is readily seen in his 2nd conclusion which contains a reference to a unit of temperature which was introduced independently from and prior to the quantification of heat by means of work. So, we have symbolically:

$$\Delta Q \underset{m}{\overset{r \,\&\, c}{\rightleftarrows}} \Delta t$$

Joule's overall approach to the mechanical equivalent of heat as expressed in the 1st and 2nd conclusions of his paper (1850) can thus be expressed symbolically as follows:

Figure 5.2: Joule's reasoning from cause to its effect, supposition about existence of causal determination, and directions measurement in the treatment of mechanical equivalent of heat

$$W \xrightarrow{r \,\&\, c \,\&\, m} \Delta Q \underset{m}{\overset{r \,\&\, c}{\rightleftarrows}} \Delta t$$

Worth noting here is that in Joule's thermodynamics, contrary to the pre-Joulian thermodynamics, the understanding of the change of heat is already separated from the understanding of the change of temperature. Each of these two understandings stands for content (form, way) of knowledge which is different from the other one. Joule's thermodynamics yields *knowledge about the cause* – change of heat due to work performed – *which is already different from the knowledge about what this cause causes* – the change of temperature due to the change of heat.

So, from the point of view the operations of measurement we face in Joule's theory of mechanical equivalent of heat as presented in his (1850) paper the situation, when a *cause is measured both by a cause in which it originates and by an effect it generates*. This I express schematically as follows (the ovals stand for the respective conceptual systems):[76]

---

76 In the diagrammatic representation of interconnection of theories by means of a shared magnitude I draw on (Pawson, 1989).

Figure 5.3: Relations between two mesuransia *sharing the same* mesurandum *in Joule's approach to the mechanical equivalent of heat*

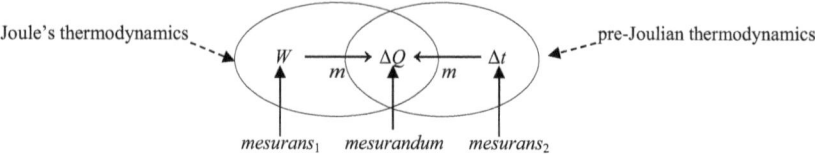

One of the lessons learned from the philosophical reconstruction by means of categories is that thermodynamics in the form which it acquired in the mid-19th century in the works of Joule – *even if in these works the quantity of change of heat was already understood as being based on the quantity of change of living force as the base* – *nevertheless did not detach itself from the method of measurement of heat by means of thermometers and their corresponding scales employing a concrete substance as they were developed in the 18th century.*

Let me now – as a conclusion of this chapter and as a test of both this lesson learned and of the efficacy of my philosophical analysis of thermodynamics of the 18th and 19th centuries by means of categories – try to determine whether these lessons also apply to Thompson's introduction of the concept of absolute temperature based on an absolute thermometric scale.

5.3 A Test: Thomson's Concept of Absolute Temperature

Thomson presented the idea of an absolute thermometric scale and thus of the concept of absolute temperature based upon it as follows (1848, 313–314):

> The theory of thermometer is ... as yet far from being in [a ...] satisfactory state. The principle to be followed in constructing a thermometric scale might at first sight seem to be obvious, as it might appear that a perfect thermometer would indicate equal additions of heat, as corresponding to equal elevations of temperature, estimated by the numbered divisions of its scale. It is however now recognized (from the variations in the specific heats of bodies) as an experimentally demonstrated fact that thermometry under this condition is impossible, and we are left without any principle on which to found an absolute thermometric scale.

In the subsequent research, by drawing both on Carnot's (1824) and the works of Joule, he arrived in cooperation with the latter at the statement which they view as an explicit definition (Joule – Thomson 1854, 351):[77]

---

[77] On this see his works (1849; 1853a; 1853b; 1853c). For their analysis see (Cardwell 1971), (Forrester 1975), (Smith 1977), (Cropper 1987) and (Chang and Yi 2005).

If any substance whatever, subjected to a perfectly reversible cycle of operations takes in heat in a locality kept at a uniform temperature, and emits heat only in another locality kept at a uniform temperature, the temperatures of these localities are proportional to the heat taken or emitted at them in a complete cycle of the operations.

That is, it should hold = $\frac{T_1}{T_2} = \frac{Q_1}{Q_2}$. Here T_1 and T_2 are values of the magnitude called "absolute temperature," in the sense that they assigned to it from an absolute scale and, thus, independently from temperatures whose actual values are acquired by a thermometer employing some kind of thermometric substance.

However, right after the above given statement they declared the following (1854, 351–352):

> To fix on a unit or degree for the numerical measurement of temperature, we may either call some definite temperature, such as that of melting ice, unity, or any number we please; or we may choose two definite temperatures, such as that of melting ice and that of saturated vapour of water under the pressure 29.9218 inches of mercury in the latitude 45°, and call the difference of these temperatures any number we please, 100 for instance. The latter assumption is the only that can be made conveniently in the present state of science, on account of the necessity of retaining a connection with the practical thermometry as hitherto practiced; but the former is far preferable in the abstract, and must be adopted ultimately.

A similar situation was already encountered in Thompson's paper (1848). At the level of abstract theory he presented his ultimate aim as follows (1848, 316):

> The characteristic property of the scale which I now propose is, that all degrees have the same value; that is, that a unit of heat descending from a body A at temperature T° of this scale, to a body B at the temperature (T-1)°, would give out the same mechanical effect, whatever the number T. This may justly be termed an absolute scale, since its characteristic is quite independent of the physical properties of any specific substance.

But, then, again like in the paper (Joule – Thomson 1854), he – by taking into account the actual practice of measurement of temperature by means of a thermometer with a specific substance, namely, air – stated that "The unit of heat adopted is the quantity necessary to elevate the temperature of a kilogram of water from 0° to 1° of the air-thermometer" (1848, 317).

So, both Joule and Thompson in fact acknowledged that even if in the relation $\frac{T_1}{T_2} = \frac{Q_1}{Q_2}$ the ratios $\frac{T_1}{T_2}$ and $\frac{Q_1}{Q_2}$ have, with respect to each other, the status of a *mesurandum* and *mesurans* respectively, still in the practice of measurement the actual values of ratio $\frac{Q_1}{Q_2}$ have to be determined by means of $\frac{t_1}{t_2}$, that is, by ratios of temperatures tied to a thermometer with a specific substance. So, the ratio $\frac{Q_1}{Q_2}$ stands in relation to — as a *mesurandum* to a *mesurans*.

Drawing on the above used symbols, the measurement situation – with respect to the concepts of temperature, heat and absolute temperature – as given in the Joule-Thomson thermodynamics in the mid-19th century can be represented as follows:

Figure 5.4: *Relations between temperature, heat and absolute temperature as* mesurans *and* mesurandum *in Joule-Thomson thermodynamics in mid-19th century*

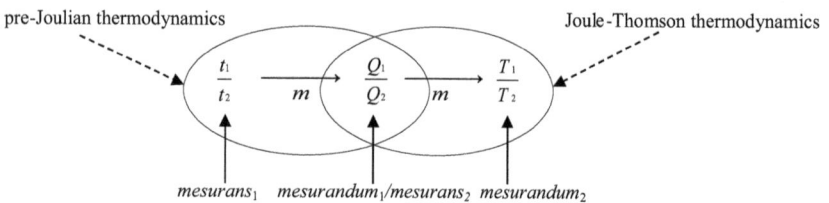

Part II:
Methodological Issues in Primate Research

Chapter 6: The Methodological Turn in Ape Research: Sue Savage Rumbaugh

The aim of this chapter is to investigate – from the point of view of philosophy of science and philosophy of social science – the turn in the Ape Language Project (hereafter, ALP) as accomplished in the works of Sue Savage-Rumbaugh and her collaborators from the seventies of the 20th century till the first decade of the 21st century. In this project took place a highly interesting turn from the orientation of research on natural sciences to that on humanities. I shall analyze all the relevant works of Savage-Rumbaugh *et alia* from the point of view of the three central levels of ALP: its *scientific* level, its *metascientific* level where the central concepts of ALP are reflected upon, and the *methodological* level (as a consequence of certain conceptual choices at the metascientific level).[78]

I shall present, first, the scientific level of the ALP. Then I shall deal with the metascientific level and show how certain conceptual choices in this level lead to certain preferences in the methodological level, that is, to certain choices of applied methods and, finally, to a shift of the ALP to the realm of the humanities.

The impetus to deal with the ALP came to me from Peter Koťátko's article (2005) located in the framework of philosophy of language, where he stated the following (Koťátko 2005, 13–15):

> For a naïve realist, language is one of the parameters of human existence in the world: a tool of communication understood as a process taking its course in the real world. ... If the reader can bear my quasi-Hegelian vocabulary, language communication and language as its means are one of the products of human phylogeny which is part of the self-development and self-differentiation of the world. ... Let us return to the naïve-realist explanation of language as a product of human phylogeny. The naïve realist starts with the presupposition that the ancestor of man is, like other living beings, situated in the real world by such parameters of his existence like the type of exchange with the environment, [sensory-]motor apparatus, the nature of sense receptors, schemes of perception, patterns of instinctive and acquired behavior. All this codetermines what elements

---

78 Here I shall not report on, with few exceptions, the quantitative results of the experiments of Savage-Rumbaugh and her collaborators. The measure of a successful experiment was set by her at 90% or better, while 50% and worse were viewed as pure chance. Results in between were viewed with suspicion leading to repeated instructions and retesting of the apes or to a reorganizations of the experiments. The experiments were composed of several training sessions and then of test trials where completely novel entities, in respect to training, were introduced.

or features of the environment, what differences and similarities can be grasped by him ... what is an object for him, how he orients himself in the world. Accordingly, one has here already – before the becoming of language – a rich differentiated picture of the world. ... As to the degree of this differentiation one has no reason to withhold to the ancestor of man anything we usually ascribe to other primates.

It is worth noting here that even if Koťátko begins from a purely *a priori* standpoint in presenting his concepts pertaining to language, he then makes two claims that are of an *a posteriori* nature: one about the ancestor of *Homo sapiens* and one about nonhuman primates. While I leave it in this chapter open if his claim about the ancestor of *homo sapiens* can be empirically tested, his claim about nonhuman primates leads directly to the following question: *What is the relation between, on the one hand, an a priori conceptual framework presenting certain views about language and, on the other hand, the implications of this framework in the ALP at the level of empirical research, theory formation, and methodology of this formation?* This chapter shall try to answer this question.

6.1 The Starting Points: The Late Sixties, Early Seventies, and the Lana Project

The attempts to use *nonvocal* ways of language communication with apes came under way after attempts at teaching ape human vocalization failed.[79] In addition, it became clear,[80] that the apes involved in experiments in that time were not able to control voluntarily their exhalation and thus to produce all those phonemic sounds humans are able to produce.[81] This led then to language projects in the framework of which the apes should be taught language by using specific symbolic systems, for example, American Sign Language (ASL), as in the endeavor of Allen and Beatrice Gardner with the chimpanzee (*Pan troglodytes*) named Washoe.[82]

Already here it is noteworthy that Gardners' endeavor was in its structure and course driven by the supposition that the main method of introducing new signs to Washoe should be *imitation*: "As a method of prompting, we have been able to use imitation extensively to increase the frequency and refine the forms of signs" (Gardner – Gardner 1969, 666). This imitation, which they viewed as an important

---

79 On this see (Hayes and Hayes 1951).
80 On this see, for example, (Lieberman 1968).
81 For more recent results see (Taglialatela and Savage-Rumbaugh and Baker 2003), (Savage-Rumbaugh and Fields and Spircu 2004), and (Benson *et al.* 2004).
82 On this see, e.g., (Gardner and Gardner 1969) and (Gardner and Gardner 1971).

principle of language acquisition both in human infants and apes, should have as its basis, they claimed, *instrumental conditioning* (Gardner – Gardner 1969, 668). And they claimed also (Gardner – Gardner 1971, 129):[83]

> The acquisition of individual signs is the aspect of this project that is most clearly related to the paradigm of S-R reinforcement theory. This paradigm, which had a strong influence on the tactics that we used for teaching individual signs, serves as a convenient point of departure for the description of specific teaching methods. According to this theory, particular responses are made in particular stimulus situations. When a response is followed promptly by reward, there is an increment in the probability that that particular response will be repeated by the subject when the particular stimulus situation is repeated. Thus, the first step in training is to have the subject make the to-be-learned response in an appropriate situation.

That basis was initially accepted also in the project initiated by D. M. Rumbaugh which involved a chimpanzee named Lana (Gill and Rumbaugh 1977, 155, 157, 158).

However, the Lana project displayed, when compared with the Washoe project, two important differences. First, instead of the signs of the ASL, the medium of communication was purely visuographic symbols without any vocalization, so-called *lexigrams*, designed of basic 9 elements which could be combined into complex symbols,[84] and which were implemented on a computer keyboard.

By selecting and depressing particular keys on which lexigrams were embossed, Lana could actuate a vending device, controlled by the computer, to provide her with a particular food, for example, M & M, a piece of banana, a sweet potato, etc.

Second, in the project started by D. M. Rumbaugh, Lana was initiated into language not by teaching her from the outset particular lexigrams, but by teaching her whole "stock sentences." Initially, the lexigrams of a stock sentence, say "Please machine give M & M.", were interconnected in such a way that the depression of just one lexigram lighted all the keys and arranged them into the right order, which led to a dispensing of the respective food. The whole stock sentence was then gradually split up as follows. First, the lexigram for the requested entity was separated and Lana had to depress two keys: "Please machine give" and "M & M." Then, she had to press three keys: "Please machine", "give", and "M & Ms.", next, four keys: "Please, "machine", "give", and "M & M." Then,

---

83 Here "S-R" stands for "stimulus-response."
84 On this see (Rumbaugh *et al.* 1973, 386) and (Savage-Rumbaugh and Rumbaugh 1978, 271).

five keys had to be depressed: "Please", "machine", "give", "M & M", and "."[85] Next, the keys were randomly distributed among a larger number of keys and, finally, the location of the keys on the keyboard was regularly changed.

The ALP with Lana also involved a gradual enlargement of the vocabulary taught to her in such a way that new names of food items to be requested were introduced as well as new types of requests that could be addressed to the machine to provide either a food-type entity or a drink-type entity. Also, Lana was taught to state requests, introduced by the "?" lexigram, addressed to an experimenter, e.g., "?You put chow into machine".

The *results* of the Lana project were twofold. *On the one hand*, Lana manifested skills of learning new lexigrams-words as well as the ability to hold to *syntactical* rules of stringing lexigrams into sentences. Thus, when facing a set of three stock sentences – two syntactically correct and one incorrect – she was able to eliminate the syntactically[86] defective sentence with a success rate of 90%. And she was able to complete stock sentences stopped by the human experimenter at any lexigrams. She also displayed the ability to apply previously learned sentences to novel situations. So, for example, when being purportedly misinformed by the experimenter about the content of a dispenser (e.g., chow was put in, but the experimenter claimed that he put in a more preferred food), Lana after several failed attempts to obtain it stated her doubts about the experimenter's claim and then asked for the more preferred food to be put in the dispenser. Finally, when a new type of food she was not able to tag by a lexigram was put into the dispenser, she asked for the *name* of the food instead of the food itself.[87] This means that Lana was employing lexigrams in two different ways: to refer to an extra-linguistic entity and as a meta-language expression referring to an object-language expression.

On the other hand, the results of the Lana project, together with the results of the Washoe project subject, were subjected to severe criticism by Savage-Rumbaugh, which served as the starting point for the development of the ALP. The main points of this criticism were as follows:

1) Because at the beginning of the ALP its aim was to find out if a chimpanzee would "learn a synthetic language, and would … evidence sensitivity to the rules of word combination written for that language" (Rumbaugh – Savage-Rumbaugh 1978, 124), the whole Lana project was limited to the realm of

---

85   Here the lexigram "Please" signaled to the computer software an impeding request and the lexigram "." (dot) its termination.
86   Defective was either the order of lexigrams or a word was added in excess.
87   On this see (Gill 1977).

syntax, like the Washoe project. And, although the symbols (lexigrams or signs of ASL) utilized were supposed to function referentially (i.e., semantically) for the apes, the semantic skills of Lana and Washoe had not, at that time, been properly tested.

2) Because in the Lana project the intent was "to teach Lana language by differentially selecting and shaping, through Skinner's operant conditioning procedures, her responses at and to the keyboard" (Rumbaugh and Washburn 2003, 99), one could not exclude beforehand the possibility that (Rumbaugh and Washburn 2003, 115)

> the food or object offered might come to serve as a discriminative stimulus through operant conditioning procedures and might serve to elicit a response of pressing a specific key. Indeed, whether the animal "knows" anything about the semantic meaning of the lexigram on the key being depressed remains unknown unless other assessment are made.

This, in turn, would mean that the depressed key would be for Lana just an operant and would *lack any symbolic and semantic function for her*.

3) Lana's and Washoe's ability to label an object by producing an ASL-sign or by depressing a lexigram-key should not be interpreted as the ability to use the symbol referentially. What could, in fact, be at work here is just "a simple association between label and object, and the ability to label is not sufficient criterion for a symbol to be viewed as a word or as an element of a chimpanzee's symbolic vocabulary" (Savage-Rumbaugh and Rumbaugh and Boysen 1980, 52).

4) In the case of the results of experiments reported by the Gardners, one could not properly distinguish "whether the chimpanzee is simply imitating or echoing in a performative sense, the action, or object, or whether the animal is … attempting to relay a symbolic message. The high degree of iconicity in ASL … could permit the animal to recall the proper hand motion simply by observing either the object or the form of the action" (Savage-Rumbaugh and Rumbaugh and Boysen 1978b, 550).

5) The experiments with Washoe and Lana were based on the wrong presupposition that once the ape was taught to *produce* a symbol, he would also understand its meaning, that is, grasp its *semantics*, and its use to convey symbolic information to another subject about an object, that is, come to terms with its *pragmatics*. However, when Lana was presented not with the objects themselves, but with the *symbols* of these objects, she was not able to deal successfully with these symbols (Savage-Rumbaugh and Rumbaugh and Boysen 1980, 49–60; Savage-Rumbaugh and Rumbaugh and Fields 2006, 522).

From points 1) through 5) Savage-Rumbaugh drew the conclusion: "We can find no definite demonstration that Washoe ... [and] Lana ... used symbols representationally" (Savage-Rumbaugh and Rumbaugh and Boysen 1980, 55), and that "there is no evidence that [they] achieved symbolization proper" (Savage-Rumbaugh and Rumbaugh and Boysen 1980, 60).

Accordingly, Savage-Rumbaugh moved to more advanced experiments with the chimpanzees Sherman and Austin to try to find out if apes can come to terms with the semantics and pragmatics of communication.

6.2 The Sherman-Austin Project

The Sherman-Austin project started with a negative result in a series of experiments where these two apes, together with two other apes (Kenton and Erika), were taught to depress keys corresponding to one of several present objects A, B, C, etc., which were present simultaneously and which, even in case of a successful choice of a lexigram, they would not obtain.[88] And even if they were able to assign to each of these objects its correct lexigram, when just one object was present, once the number of objects reached the threshold of three, Sherman's and Austin's performance dropped significantly.

It soon became clear to the researchers that while the human experimenter wanted the chimps to attend to the relation of the displayed object to its corresponding lexigram, the chimps attended to another relation in the sequence of events, namely, the relation of the key to be depressed and the object (food) they would receive as a result of depressing it. Thus, *in the absence of training*, the apes involved were not able to attend to the whole *practico-linguistic* complex composed of (i) action on object in the world (the experimenter picks up an object and shows it to the ape), (ii) the relation of that object to the corresponding lexigram (the ape depresses a lexigram-key which corresponds to the object), (iii) the intersymbolic relation (the experimenter communicates to the ape words of praise for their correct choice or, in case of a failure, points to the lexigram the ape should have depressed), and (iv) the renewed action on objects in the world (the ape is rewarded/not rewarded).

In another series of experiments, aimed at finding a remedy to the failure of the previous series of experiments, the apes were split into two subgroups: the event group (Sherman and Kenton) and the label group (Austin and Erika). The difference between these two groups was in the activities they were allowed and

---

88 On this see (Savage-Rumbaugh 1986, 61–67).

required to perform. The event group was subjected to an *active* paradigm while the label group to the *naming* paradigm. In the former, after finding in several sessions their members' order of preferences for food types (the order was M & M, sweet potato, chow, juice), the dispensers were loaded with foods/drinks of these types, and they were then tested for their ability to assign to them, while asking the machine for dispensing the respective food, the corresponding lexigram. Initially, only their ability to name just M & Ms was tested with only the M & M-lexigram lighted on the keyboard.

After this test was mastered in an errorless way, an additional and irrelevant key (whose depression did not yield an M & M) was added to the keyboard. Once the ape was again able to choose the M & M-lexigram with a 90% rate of success, a third key, again irrelevant, was made operational on the keyboard.

In the label group the task was initially to assign the M & M-lexigram to the M & M held up by the experimenter; here also the number of irrelevant keys was gradually increased. In case of just one food type – M & M – the results in both groups were very similar. But, once additional lexigrams were introduced, the results were profoundly different. The apes in the event group learned very rapidly the names of all four food types introduced, while in the label group one ape learned to label only two types of food while the other just one. In fact, this latter result held only in the case when the *condition of difference of preference* was at work and the apes from the label group had to choose between two lexigrams for food where one food was much more preferred than the other. But, once that condition was cancelled in such a way that the apes had to choose from two food types for which they had approximately the same order of preference (say, M & M and banana, or, chow and sweet potato), then the apes from the label group failed while those from the event group succeeded.

From the point of view of *theory*, the differences between the event and label groups were interpreted by Savage-Rumbaugh to mean that the apes from the label group simply associated one symbol with a more preferred food and another symbol with a less preferred food. But, they failed to do this once the requirement was to perform the *purely referential operation* of assigning a symbol to a particular food when there were several symbols to choose from. In the case of the event group, its members were stimulated to take an active approach with respect to their environment – to bring the machine to vend a preferred type of food – by choosing the respective symbol. That encouraged the recognition of the relation of the lexigram to the corresponding food item. However, Savage-Rumbaugh concluded that his relation was *not* recognized by the apes from the event group as one of *reference* because they displayed the following unusual behavior.

They were able to shift in their key-depressing behavior from a lexigram for the more preferred food to a lexigram for a less preferred food once the dispenser – placed in full sight of the apes – ran out of the former *and* its corresponding key was darkened by the experimenter. But, once the dispenser ran out of the more preferred food *and* the corresponding key was left illuminated by the experimenter, the apes persistently continued to press this key even after the machine had already stopped vending the food corresponding to this key. From this odd behavior Savage-Rumbaugh drew the conclusion that at this level of instruction a purely referential capacity (i.e., a one-to-one correspondence between a lexigram and its reference) was still not attained in the apes even from those in the event group. Instead, (1986, 74)

> their previous training had enabled Sherman and Kenton to encode food desires in a primitive way. ... At this point they employed symbols only in a primitive cause-effect manner—just as simple actions like pushing, shoving, or biting can be used to affect another directly—rather than to communicate.

In order to attain in the chimpanzees a referential capacity, a set of experiments with an intricate structure was devised. Initially, the apes were taught to relate the vending of food with the state of the dispenser in such a way that only one dispenser was loaded – in full view of the apes – so that depressing the correct keys (say, "Please machine give beancake.") yielded the requested food. At the same time, the dispenser was connected to the computer in such a way that once the correct sequence of keys/lexigrams was depressed, the dispenser started to rotate as though it was going to vend the requested food.

When the apes acquired this capability, the number of dispensers was increased to two, one located to the right and the other to the left of the keyboard, and with the key for the food-type in the left dispenser placed on the left side of the keyboard and the key for the food-type in the right dispenser placed on the right side of the keyboard. In this way the apes could coordinate more easily the type of food in a dispenser with its corresponding lexigram. Initially, *both* dispensers were filled with the respective foods and once the chimpanzees mastered the naming task at this level, only one dispenser was filled with food per day, while the other remained empty during the same period of time. So, for example, if one dispenser was filled with pieces of banana and the other was empty, depressing the keys in the sentence "Please machine give beancake" only caused the rotation of the dispenser but no vending.

While till now the task was to teach the apes to correlate a *same-side lexigram with a food-in-a rotating (same-side) dispenser*, once this task was mastered, this complexity was gradually reduced. First, the respective lexigrams were removed from their side-location on the keyboard and randomly distributed on the keyboard, and the apes were then subjected to a series of tests. Once they mastered the correlations in this new arrangement – *lexigram and food in two rotating dispensers* – the number of dispensers was reduced to one, thus eliminating the dispenser factor and reducing the whole experiment exclusively to the semantics of the complex *lexigram-food-type*. After prolonged training the apes were successful in mastering these semantics at a rate of 90% or higher.

Savage-Rumbaugh then designed another type of training and testing regimen, also with the aim of investigating the apes' semantic capacity to relate symbols and their referents, the latter now *not present*. In the training phase, Sherman and Austin were, first, trained to label three inedible items (stick, key, money). Next, they were taught to sort six items (orange, bread, beancake, stick, key, money) into two categories, food and tool, via the "edible vs. inedible" functional distinction by placing them into two separate bins: one for foods and one for tools. Once this task was mastered, separate lexigrams for tool and food in general were introduced in addition to the bins with their particular objects. Once the correspondence in the complex triple objects-bin-lexigram was mastered by the apes, the complex was reduced by removing the bins, and the apes then had to label each of the six objects either with the lexigram "food" or the lexigram "tool" (i.e., they had to create pairs instead of triples). After this level of training was successfully mastered, photographs of the six objects mentioned above were introduced, initially by taping them to the objects themselves, while the apes were asked to assign the lexigram "food" or "tool" to the respective pairs. Once this task was mastered, the objects were removed and the apes had to label just the photographs with one of the two available lexigrams. After this task was mastered, the last and crucial phase of the experiment was initiated. The task was to assign the tool-lexigram or the food-lexigram to the *lexigram* for each of the six objects. Initially, the tool-lexigram and food-lexigram had to be assigned to the photograph-lexigram pair and then, after this was mastered, the photographs were removed. Finally, new items (5 food-items and 5 tool-items) were introduced.[89] Figure 6.1 reproduces the results obtained by Sherman and Austin together with an iconic representation of the sequences of training and testing the apes underwent.[90]

---

89 Both apes were able to assign to each of these items its lexigram but they were not taught to make a categorical assessment of the lexigrams or of the very objects corresponding to them.
90 On this see (Savage-Rumbaugh 1981, 46–47). The left column reproduces the training sequence, the right the testing sequences. In order not to overburden the figure, I eliminated the results for Lana.

Figure 6.1: Flow chart of experiment with items learned in the left column and items tested in the right one

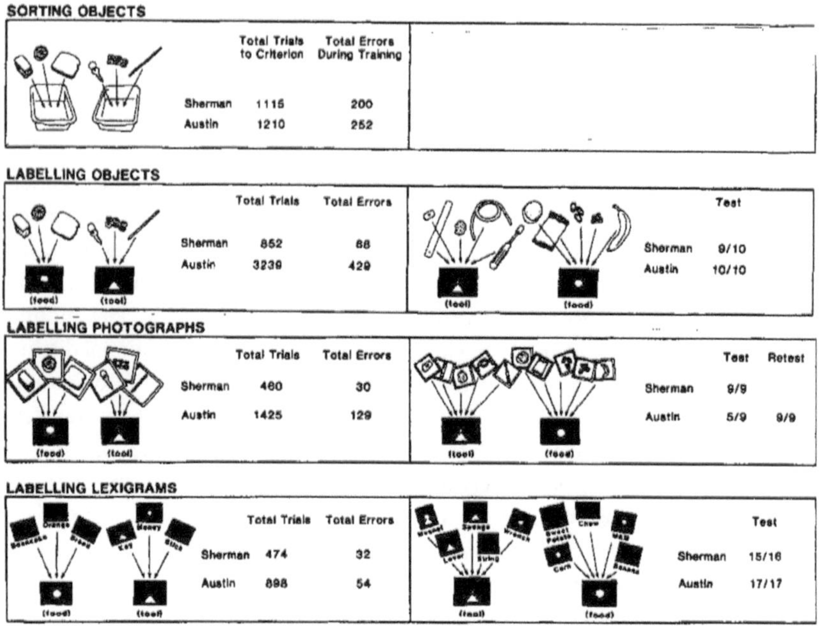

SORTING OBJECTS

	Total Trials to Criterion	Total Errors During Training
Sherman	1115	200
Austin	1210	252

LABELLING OBJECTS

	Total Trials	Total Errors		Test
Sherman	852	88	Sherman	9/10
Austin	3239	429	Austin	10/10

(food) (tool)   (tool) (food)

LABELLING PHOTOGRAPHS

	Total Trials	Total Errors		Test	Retest
Sherman	460	30	Sherman	9/9	
Austin	1425	129	Austin	5/9	9/9

(food) (tool)   (tool) (food)

LABELLING LEXIGRAMS

	Total Trials	Total Errors		Test
Sherman	474	32	Sherman	15/16
Austin	898	54	Austin	17/17

(food) (tool)   (tool) (food)

All experiments mentioned above were testing for Sherman's and Austin's (possible) *semantic* capacities and skills. Savage-Rumbaugh, however, proposed an additional series of tests aiming at their *pragmatico-linguistic* skills and capacities, that is, tests for "behavior characteristic of true speech episodes in which a listener and a speaker use symbols to control and coordinate each other's behavior in meaningful rule-bound exchanges" (Savage-Rumbaugh 1986, 113).

Initially, the experiment involved only one ape (hereafter, A_1) and only one human experimenter (hereafter, E_1) located in different rooms, each having a keyboard at his disposal. The experimenter baited a container with food from a refrigerator, while being watched by the ape who, however, did not know what particular food was put into the container because his vision was blocked by the door of the refrigerator. Then, the experimenter informed the ape about the content of the container by depressing a key on his keyboard. The ape, once being thus informed, in turn depressed the corresponding lexigram on his own keyboard, and in cases of a match of the lexigram with the food from the container, was rewarded by the experimenter.

In a subsequent experiment, the number of apes was increased to two, as was the number of experimenters, while just one keyboard was used initially. Ape A_1 accompanied experimenter E_1 while he was baiting the container with food. Then A_1 carried the container to experimenter E_2, who did not know what type of food was baited and accompanied A_1 to the keyboard, where A_2 was waiting. A_1 then informed A_2 via the keyboard about the content of the container. A_2 then responded to this information by depressing the same lexigram on the same keyboard; in case of a match of this symbol with the content of the container, both apes were rewarded with the food in the container. Next, the uninformed ape was tested for (possible) comprehension of information he obtained from the informed ape as follows: three photographs were placed in front of him and the task was to point to the one which referred to the content of the container.

Finally, the whole experimental structure was configured in such a ways as to exclude the possibility that the uninformed chimp could simply imitate the informed one by depressing the same key. Accordingly, the chimpanzees were placed in different rooms with each having a separate keyboard at his disposal, and the same symbols were placed in different locations on these keyboards.

In all experiments just mentioned the chimpanzees achieved a high level of success as can be seen from Table 6.1 (Savage-Rumbaugh 1986, 140).

Table 6.1: Sherman's and Austin's successful performance on the first interanimal communication task

Experimental condition	Trials (correct/total)	Proportion correct
First interanimal	test	
Animals use single keyboard. Experimenter knowledgeable (Vocabulary size = 36)	33/35	.9
Control conditions		
Animals use single keyboard. Experimenter blind	60/62	1.0
Animals use separate keyboards, observe one another through window. Experimenters blind. (Vocabulary size = 40)	36/40	.9
Animals use single keyboard, observing animal pointing to photograph of food following his request. Experimenters blind.	27/30	.9

These experiments with Sherman and Austin were followed by a sequence of new experiments designed to test them in a communicative situation that was not controlled by a human experimenter, as before, but by the chimpanzees themselves. Noteworthy also is the *novelty* of these experiments, as compared to the earlier ones, because the chimpanzees were not trained (i) to attend to another's statements at the keyboard nor (ii) to request food being eaten by or conversed about by another animal (Savage-Rumbaugh – Rumbaugh – Boysen 1978a, 642).

Based on the abovementioned experiments, the ape who had earlier watched the container being baited with a particular food was now able to inform the ape who did not have this information. These experiments, the results of which are given in Table 6.2 below, were conducted under the following conditions.

(1) Initially, when both apes used the same keyboard, care was taken that the receiver of the information could see the depressed key only as displayed on the screen, not the actual depressed key with its location on the keyboard.
(2) Next, a window was placed between the apes, both having now at their disposal their own keyboard. In such a way they could inform each other about the content of the container, and then ask for it.
(3) In a subsequent experiment, the ape informed by means of the keyboard was handed three photographs (one corresponding to the food in the container and two irrelevant) and the task was to name the food by means of the keyboard. Only if the chain of matches *baited food* = *lexigram depressed by the ape observing the baiting* = *lexigram depressed by the initially uninformed ape* = *baited food* held, were both apes rewarded.
(4) The final test was centered around the question whether the apes separated by a window would understand their mutual communication and on this basis share a particular food handed over to just one of them.
(5) An additional test, performed with just a few trials and thereafter terminated, is given in the bottom half of the table below. Here, the apes were to have used only vocalization and gestures. As is obvious from the results in Table 6.2, it was a complete failure (Savage-Rumbaugh – Rumbaugh – Boysen 1978a, 642).

Table 6.2: Sherman's and Austin's performances under different conditions

Experimental condition	Trials (correct/total)	Proportion correct
Animals use single keyboard but view only projected response of informed animal. (Vocabulary size = 40) Experimenters blind.	24/26	.9
Informed animal denied use of keyboard to describe contents of container. Experimenters blind.	4/26	0.2

154

What were the results obtained in the Austin-Sherman project? From the point of view of this chapter the following three have to be mentioned (Savage-Rumbaugh et al. 1993, 10–11; Rumbaugh and Savage-Rumbaugh and Washburn 1996, 117–118):

(1) Chimpanzees have a capacity for *semantics* in the sense that they can assign to symbols their referents in the extra-linguistic world, and they can understand the *meaning relations between the symbols*, even in the absence of their referents;
(2) they have a capacity for *pragmatics* in the sense that they can communicate mutually by means of symbols, once they develop skills of joint attention and if the environment puts a premium on their mutual cooperation;
(3) the semantic and pragmatic use of symbols in chimpanzees is the prerequisite for the development of their *syntactic* competencies.

Even with these positive results, the following two "self"-critiques were addressed by Savage-Rumbaugh in the framework of the Austin-Sherman project. First, like in the Lana project, even when the ability to produce symbols was already in place, Sherman and Austin required prolonged and laborious training to display receptive/comprehension skills. Second, even though Sherman and Austin were exposed to spoken English for years, they never comprehended it. The test for this comprehension was given – under the condition of screening off any possible cueing – as follows. They were presented a set of sixteen objects they were able to name by means of lexigrams, and then they were tested for the ability to pick up and hand over a particular object to the experimenter once being asked for it, first, in spoken English and, then, by means of a lexigram. As shown in Table 6.3, their ability to apply these two ways of language is worlds apart (Savage-Rumbaugh 1986, 57).

Table 6.3: Sherman's and Austin's results of the comprehension test of spoken English and lexigrams

Type of language Ape	Lexigrams		Spoken English	
Sherman	Correct item given:	16	Correct item given:	2
	Incorrect item given:	0	Incorrect item given:	14
Austin	Correct item given:	16	Correct item given:	1
	Incorrect item given:	0	Incorrect item given:	15

The reasons for that problem and this failure were left unstated in the Austin-Sherman project; a clarification and solution was given only later, in the course of the Kanzi project.

6.3 The Kanzi Project

The Kanzi project started initially as an unforeseen consequence of the classical lexigram training to which Kanzi's step-mother Matata – a so-called pygmy chimpanzee/bonobo (*Pan paniscus*) – was subjected. During her training, Kanzi accompanied Matata and was separated from Matata only after reaching the age of 30 months. At this moment, unexpectedly for all the human experimenters around him, he purportedly started to use the lexigrams from the keyboard available to him. From this the experimenters concluded that Kanzi had acquired the ability to use lexigrams just by observing and imitating Matata and not by conditional training. While in the first week of separation from Matata he used 8 lexigrams for designating even absent entities, in the 16 months that followed his vocabulary – both single words and sets of their combinations – increased substantially.[91]

This unusual behavior was caused by a profound change in the way Kanzi and the human experimenters around him interacted. Instead of the explicit training regimen to which Lana, Austin and Sherman were subjected, Kanzi was integrated into a way of life where he moved around in the woods around the language laboratory compound and had to pick up objects he initially indicated by means of lexigrams as placed in certain locations and at the same time was interacting with the experimenters both by means of a mobile keyboard as well as spoken English.

In addition to the discovery that an ape from the species *Pan paniscus* is capable of comprehending communicative symbols not by training but by social integration, the experiments with Kanzi led to the following conclusions:

(1) This species of ape is capable of comprehension of spoken and synthesized English.
(2) He acquires lexigrams not only piecemeal but also in sentential complexes.
(3) He is capable of passing the test for false beliefs, that is, he has a "theory of mind"[92] in the sense that he is capable of knowing not only what is the state

---

91   For a quantification of this increase see (Savage-Rumbaugh and Rumbaugh and McDonald 1985, 659–663).
92   On this term see (Premack and Woodruff 1978a) and (Premack and Woodruff 1978b).

of affairs in the world of objects but also what another subject, say a human experimenter, knows about that state of affairs, even if they mutually differ.
(4) He is capable of producing stone tools, and he acquired this skill not by means of conditional training but by looking at another subject performing this activity.

Let us deal with some of these results.[93]

1. *Comprehension of Spoken English* (Savage-Rumbaugh 1987b, 220). The set of tests was based on an experimental complex set up by Kanzi, two human experimenters, a test booklet with three lexigram symbols in it, and the task to test Kanzi. In order to preclude beforehand any possible cueing, one of the experimenters arranged the booklet, pronounced an English word, and then passed the booklet to another experimenter, who handed it to Kanzi. Kanzi then opened it in such a way that only he could see the target item in it. During the choice, the second experimenter noticed the location of Kanzi's arm on the booklet and then he checked if Kanzi selected the lexigram corresponding to the spoken word. In the test, Kanzi was presented with 194 spoken words; he was 100% successful with 109 words, 75% successful with 40 words, and below chance with the remaining 45 words – the interpretation of the last result was that these words were as yet not acquired by him.

The initial experimental complex experimenter$_1$-experimenter$_2$-Kanzi was then enlarged in order to involve Austin and Kanzi, as well. Even though all three apes were subjected to the same use of English by the experimenters, their comprehension of spoken English varied substantially, as can be seen from Table 6.4 (Savage-Rumbaugh *et al.* 1985, 184).

Table 6.4: *Comparison of Kanzi's, Sherman's and Austin's test results for comprehension of lexigrams and spoken English*

	lexigram		spoken English	
	total score	percentage correct	total score	percentage correct
Kanzi	88/93	95%	83/93	89%
Sherman	92/93	99%	47/93	50%
Austin	91/93	98%	55/93	59%

---

93 For the results of tests for comprehension of synthesized speech see (Savage-Rumbaugh 1987b, 222, 228–229); for the production of stone tools see (Schick and Toth and Garufi 1999), (Toth *et al.* 1993), (Toth and Schick 2006), and (Toth and Schick 2009).

The types of entities involved in this type of experiment were then enlarged further (Savage-Rumbaugh et al. 1986). In addition to lexigrams, spoken English words, and their referents, *photographs* and *synthesized speech* were also introduced. This, in turn, enabled them to create, in addition to the original test pairs (i.e., *spoken English → referent* and *lexigram → referent*), the following new test pairs: (i) photograph → lexigram, (ii) spoken English → photograph, (iii) spoken English → lexigram, and (iv) synthesized speech → lexigram. In the case of (i) the ape was shown a photograph and then asked to select, from a set of three alternatives, the corresponding lexigram; in the case of (ii) the task was to assign a spoken English word, taken from a set of three alternative words, the appropriate photograph; in the case of (iii) the ape had to select the lexigram corresponding to a spoken word; finally, in (iv) a lexigram had to be selected corresponding to a word produced by a speech system.

Like in the previous tests where English words were involved, Kanzi was performing substantially better than Sherman and Austin.[94] The propensity of apes of the *Pan paniscus* species to acquire symbol comprehension thus seemed to be markedly higher than that of apes from the species *Pan troglodytes*. There are at least two possible explanations: the basis of this difference is the difference given universally between these species and thus having the origin exclusively in *phylogeny*, or it is given exclusively at the level of *ontogeny* of the particular apes involved in the experiments. Savage-Rumbaugh, in order to test these competing hypotheses, created an experimental set up where the possible *differences in ontogeny* between a bonobo, on the one hand, and a chimpanzee, on the other, were eliminated by subjecting both to the same socializing conditions starting from the same age. This was accomplished by simultaneously teaching language skills to two apes of nearly same age – Panzee (*Pan troglodytes*) and Panbanisha (*Pan paniscus*) – from early infant age by means of observation. This enabled then to show the similarities as well differences[95] between the *Pan* species. From the point of view of this chapter the most important convergence is that both displayed *comprehension of spoken English*. Thus the mystery that surfaced in the Sherman-Austin project, namely, why these two apes had a comprehension of spoken English at the level of chance, found its natural solution. Both were introduced into an environment with spoken English too late in their ontogeny.

---

94 On the quantitative results see (Savage-Rumbaugh et al. 1986, 226–227). For a more recent of comparison of bonobos and chimpanzees see (Hermann et al. 2010).
95 For intraspecies differences in the genus *Pan* see (Sevcik and Savage-Rumbaugh 1994), (Brakke and Savage-Rumbaugh 1995), (Brakke and Savage-Rumbaugh 1996), and (Savage-Rumbaugh and Rumbaugh and Fields 2006).

Once this result was obtained, it was possible later to change the test arrangement again, now in such a way that three apes from the *same* species (*Pan troglodytes*) – Lana (age 27 years), Sherman (25 years), and Panzee (12 years) – but raised under different *paradigms*[96] – were subjected to six tests with pairs described above with two changes, namely, that, instead of a photograph, a picture generated on the screen of a computer was presented to the apes and at the same time English was spoken. The six pairs were then, for the purposes of quantification of the test results, further qualified by means of the following three categories (Beran *et al.* 1998, 181): *symbol production* for English + picture → lexigram and for picture → lexigram; *symbol comprehension* for English + lexigram → picture and lexigram → picture; and *English comprehension* for English → lexigram and for English → picture. The results of the tests showed that Panzee fared much better than the other two apes.[97] The conclusion based on these results was that in addition to the age of introduction to interaction with a language-using being, the type of interaction to which the ape is subject also plays a crucial role.

2. *Sentence Comprehension*. At the age of six, Kanzi was subjected to 310 utterances in English – each composed of two or more words – having the nature of requests. These requests were also performed by the experimenters on a keyboard, and the result quantified was Kanzi's appropriateness of response to these requests. From these 310 utterances, whose structure is given in Table 6.5, Kanzi responded appropriately to 298 (Savage-Rumbaugh 1987b, 232).[98]

---

96 Lana, as shown, above, was raised in a production paradigm, Sherman in a more interactional paradigm, while Panzee was raised in a symbolic environment from infancy, much like a human child.
97 The x-axis gives the type of category, the y-axis the number of known words.
98 A-frame stands for a location in the woods where a hut with an A-shaped roof was located; childside is the location in the laboratory where research with children was carried out. "Would you like to ball chase?" stands for a situation when Kanzi picks a ball, goes with it to the keyboard and comments "chase."

Table 6.5: Utterance types and their examples addressed to Kanzi

Sentence Types Directed to Kanzi	Examples
Action-Object	Would you please carry the straw?
Object-Action	Would you like to ball chase?
Object-Location	Would you put some grapes into the pool?
Action-Location	Let's chase to the A-frame.
Action-Recipient	Kanzi, chase Kelly.
Action-Object-Location	I hid the surprise by my foot.
Object-Action-Location	Kanzi, the pinecone goes in your shirt.
Action-Location-Object	Go to the refrigerator and get a tomato.
Agent-Action-Object	Jeannine hid the pine needles in her shirt.
Action-Object-Recipient	Kanzi, please carry the cooler to Penny.
Action-Recipient-Object	Go play with the dogs on the childside.

Once Kanzi's ability to comprehend complex sentences surfaced, Savage-Rumbaugh brought in another subject, namely, a human infant named Alia (between 18 and 24 months of age during the test period). The aim here was to compare Kanzi's (at the age of 8 years) and Alia's abilities to comprehend novel English sentences with varying lexical units and syntactic construction, where both from an early age were exposed to both spoken English and lexigrams. Due to these commonalities of treatment (Savage-Rumbaugh et al. 1993, 24):

> both species respond[ed] by learning how to (a) decode sounds into word units, (b) map these word units onto real-world cause-and-effect relations, (c) reconstruct the rules governing the combinatorial usages of different classes of these word units, and (d) use these relations and units in a productive manner to change the behavior of others so as to suit their own interests.

Savage-Rumbaugh and her collaborators then concluded the following (Savage-Rumbaugh et al. 1993, 98):

> The clear outcome from the present study is that two normal individuals of different ages and different genera (*Homo* and *Pan*) were remarkably closely matched in their ability to understand spoken language. A 2-year-old human female and an 8-year-old bonobo male demonstrated that, under relatively similar rearing circumstances and virtually identical test conditions, they could comprehend both the semantics and the syntactical structure of quite unusual English sentences.

3. *False-belief test.* This type of test[99] aimed at finding out if an ape who already had exposure both to spoken language and lexigrams was able to *understand the state of mind of others* – their intentions and knowledge – while this understanding is located in a process of mutual conversation, and – at the same time – *differentiate between the state of mind of others about the state of affairs in the world outside the realm of mind and his/her own state of mind and the state of affairs in the world.* This type of test by very its structure set the experimental complex in which the test had to take place. It involved, in addition to the ape, two experimenters, a container, and two types of food (e.g., candies and pine needles). Initially, one experimenter asked the other for candies; the latter complied with his/her request by putting candies into the container. Then, for a period of time, the first experimenter left the room where the experiment took place. During this time the second experimenter replaced, under the eyes of the ape, the candies with pine needles. The first experimenter then returned and declared that she wanted some candies from the container. She was handed the container and pretended that she could not open it. Then, finally, the ape was asked "What the experimenter *x* is looking for?" Even though the ape being tested, Panbanisha, was not subject to any prior special training, she was able to master the false-belief test. Thus, she demonstrated that in addition to the ability to decode the *syntax* of the sentence of the type "What the experimenter *x* is looking for?", she also had the ability to understand the *semantics* of this sentence – namely, that it pertains *not* to the actual content of the container but to the actual state of mind of another being as related to the content of container – and the ability to differentiate between them. Finally, she demonstrated the ability to grasp the *pragmatic* dimension of the sentence, namely, that the subjects involved in communication and in the actual test have a different knowledge about the state of affairs in the world.

The results of the Kanzi project can be summarized as follows (Rumbaugh – Savage-Rumbaugh – Washburn 1996, 119; Savage-Rumbaugh and Shanker and Taylor 1998, 207; Savage-Rumbaugh 1987a, 289):

(1) In both bonobos and chimpanzees, language skills appear spontaneously without formal training in the following order: first, comprehension of spoken words, then, comprehension of lexigram symbols and, finally, productive use of lexigrams.
(2) These comprehension skills involve the ability to understand novel words as well as sentences.

---

99 On this test see (Savage-Rumbaugh 1997), (Call and Tomasello 1999), (Krachun et al. 2009; 2010) and (Tomasello and Moll 2013).

(3) They can learn to differentiate English phonemes and can understand their combination to be words.
(4) They know the written symbol that corresponds to many of the spoken words, and they can use this symbol.
(5) They have not produced any speech that is interpretable as English words.
(6) They have not progressed in the development of language skills at the rapid space displayed by human beings.

6.4 Metascience and Methodology: From Behaviorism to Narrative Ethnography

Let us now move from the level of science of the ALP to that of *metascience* and *methodology* involved in ALP, because only by their analysis can one understand the basis and framework of the methods employed in an empirical theory.

At the level of meta-science, the central concepts enabling one to frame the language complex were given by Savage-Rumbaugh and her collaborators; and they are as follows.

A. *Language* as a process in use organizes the behavior between individuals and promotes inter-individual systems of behavior (Savage-Rumbaugh 1990, 68) or, stated otherwise, "To use language is to engage in the adaptive behavior of altering one's social environment by talking to others" (Savage-Rumbaugh – Brakke 1990, 314). Language at the same time can be viewed as a way of *intentional communication* characterized by both *reference* and *acts of referring*. "Reference" should stand for the "descriptive term for the process of using words to achieve coordinated actions towards objects, locations, other persons, and so on" (Savage-Rumbaugh – Brakke 1990, 316) and "is an inter-individual process in which a symbol (or a group of symbols) is employed for the purpose of causing another party to think or behave in a specific way" (Savage-Rumbaugh – Brakke 1990, 316). On the other hand, "referring" stands for the usage of an intentional symbolic act to accomplish certain specific goals (Savage-Rumbaugh – Brakke 1990, 316).

B. What has been stated till now about the language complex is based on the existence and use of *symbols* in the interaction of the speakers and listeners. In this interaction (i.e., communication) the relation of the symbol to its *referent* is both produced and comprehended in such a way that the symbol can be substituted for its referent; the symbol and its referent are mutually separable and they are different entities (Savage-Rumbaugh and Rumbaugh 1980, 306; Savage-Rumbaugh *et al.* 1990, 223).

C. Another set of concepts employed by Savage-Rumbaugh is that of *intend, intentional communicative act,* and *awareness. Intend* should stand for entity imputed by the participant of interaction to the signals emitted by the other. That is, "they behave as if their gestures, glances and actions—when directed toward another individual—can be used to convey meaningful messages to that individual" (Savage-Rumbaugh and Scanlon and Rumbaugh 1980, 621). This type of behavior stands for a "meta-behavior" with respect to the future alterations of the behavior of others and is labeled an *intentional (informative) communicative act* (Savage-Rumbaugh et al. 1985, 178; Savage-Rumbaugh and Hopkins 1986, 310). Finally, *awareness* should stand for awareness of these intentional communicative acts (Savage-Rumbaugh and Hopkins 1986, 309) in the sense that the particular subject of awareness realizes "that the other individuals also produce verbal behaviors and that they too make choices about engaging in a behavioral response to the verbalization of others" (Savage-Rumbaugh et al. 1985, 179). And it should stand also for his/her awareness in the sense that he/she realizes that "the future behaviors of others can be a function of his own actions" (Savage-Rumbaugh and Scanlon and Rumbaugh 1980, 622).

All the terms listed above are also used by Savage-Rumbaugh in her reflections about the *syntax, semantics,* and *pragmatics* of the language complex in such a way that she shifts, as shown above, from a syntactically inspired language project – via a semantically inspired means – to one that is pragmatically inspired, not only in her experiments with apes but also in her conceptual reflections at the meta-scientific level. So, while she characterizes syntax as "an ordering of symbols according to rules" (Savage-Rumbaugh and Rumbaugh 1978, 288), the semantic dimension of language appears once apes "began to use lexigrams as referents for objects and actions around them" (Savage-Rumbaugh and Rumbaugh 1978, 288). Finally, the pragmatics of language is understood by her as mutual communication of the subjects involved and aimed at entities in the extra-linguistic sphere (Savage-Rumbaugh 1990, 66).

Savage-Rumbaugh's understanding of the pragmatics of the complex of language can thus be represented as follows (S_i and S_j stand for subjects involved in communication):[100]

---

100 I put meaning here as a mediating link, as Savage-Rumbaugh takes it into account (Savage-Rumbaugh and Brakke 1990, 316).

Figure 6.2: Savage-Rumbaugh on the pragmatics of language

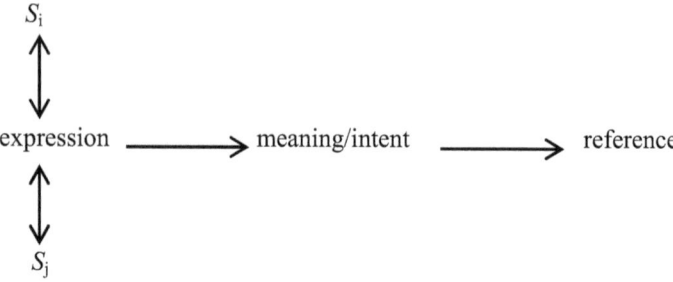

The reconstruction of the central concepts of the pragmatics of the language complex employed by Savage-Rumbaugh enables us to find the answer to the question given at the very beginning of this chapter, namely, what is the relation between, on the one hand, an *a priori* conceptual framework presenting certain views about language, and, on the other hand, the implications of this framework in the ALP. It enables us also to understand how in the framework of the ALP the respective experimental methods were developed, namely, based on reduction—as shown above—and by *variations* of the elements given in the language complex as symbolized in Figure 6.2.

In case of the Lana project targeting initially her syntactical skills, the mutual relations of lexigrams were varied by the experimenters. In the case of the Austin-Sherman project, in addition to these skills, semantic skills were also tested, in that (a) the number and types of symbols were varied with respect to pregiven reference(s) in such a way that a choice was provided from a set of symbols composed exclusively of lexigrams, or from a set composed of both lexigrams and photographs, or from a set composed of spoken English words combined with lexigrams, or exclusively from spoken English words; (b) the entities from which the apes could choose as references of pregiven symbols were varied both in number and type (e.g., food-items, inedible items); (c) the variation on the side of reference was also accomplished in such a way that a complex was experimentally constructed and then either reduced or further complicated in order to pinpoint by training and testing apes' referential capabilities; (d) apes' ability to comprehend the direction of relation in the language complex was tested by moving (i) from a reference (presented to the ape by a human experimenter) to its symbol (which was to be *produced* by the apes), or (ii) from a symbol (produced by the experimenter and communicated to the apes) to its reference (which was to be *picked up* and *presented/given* by the apes to the experimenter).

So, for example, in the experiments described above where the apes were split up into two groups: the event group and the label group, Savage-Rumbaugh applied the technique of *experimental reduction/closure*[101] (see Figure 6.3) of a complex initially composed, on the one hand, of foods of certain types to which different preferences were assigned by the apes, and, on the other hand, a set of lexigrams from which they had to choose. The reduction/closure performed stood for the replacement of those foods by foods of other types where their mutual difference in order of preference was eliminated. Thus, the apes were – by means of closure/reduction – purportedly led by the experimenters into a situation where only the *purely semantical relation of the foods to their corresponding lexigrams was given while the mediating and interfering non-semantical condition of a relation of preference to food types was eliminated.*

Figure 6.3: Operation of experimental closure/reduction

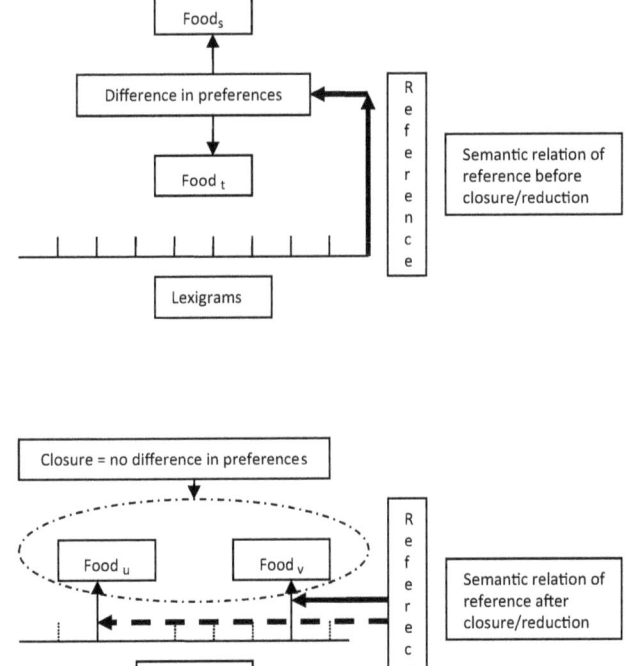

---

101 On the concept of closure see (Pawson 1989).

In the experiments with Austin and Sherman described above and involving the rotating dispensers, Savage-Rumbaugh proceeded from the methodological point of view by a *method of a successive reduction of experimentally produced complexes to their – from the point of view of the aims of the respective experiments – basic relations to be trained and then tested for their presence.*

Finally, the same method of closure was applied in the tests for spoken English to which Kanzi was subjected. Here it was useful to screen off both experimenters from Kanzi's choice of the target item, thus precluding any cueing.

In addition to the gradual shift at ALP's meta-scientific level from syntax via semantics to pragmatics, another shift happened, namely, the shift of the ALP into the realm of humanities. This shift appeared in a discussion of Savage-Rumbaugh concerning her own behavioristic approach to language, which she, at least partially, initially held.

As early as 1978 she and D. M. Rumbaugh put *cognition and generalization linked to language learning* into mutual opposition with *reinforcement and conditioning of behavior linked to language learning* as follows (Rumbaugh – Savage-Rumbaugh 1978, 120):

Cognition, the one ability absolutely essential to language, is held by the authors to be an advanced form of intellectual function that provides for the perception of relationships among the attributes of diverse things and events. ... cognition can and does result in major alterations of an organism's behavior patterns, not through the arduous selective reinforcement of certain responses at the expenses of others, but rather because of new comprehensions or understandings that come about through the emergence of perceived relationships. Based on generalized experience ... an organism becomes cognitive in its functioning.	All learning holds the potential for enhancing adaptation through behavioral alterations. ... classical and instrumental learning result in behavioral alterations through selective reinforcement of certain behaviors, at times in combinations or chains, the behaviors so "learned" are essentially basic to the response repertoire of the organism. Conditioning can and does alter the morphology of responses and the occasions for selected responses to be manifest; however, conditioning does little more than to rearrange the basic response elements of an organism's capacity and their probabilities.

Both spoke in this article in favor of the former approach and set as their aim to find out if the cognitive and generalization capabilities, once linked with language acquisition, could be traced in the apes involved in the ALP.

However, contrary to the views presented in this paper, Savage-Rumbaugh published another, later article (1984) that claimed to show that "the behavior-analytic framework, and the procedures devised to produce language-skills in apes, provide strong support for several of the major positions set forth in Skinner's (1957) *Verbal Behavior*" (Savage-Rumbaugh 1984, 223). However, a careful analysis of this article brings to the surface rather ambiguous results. On the one hand, she shows how the arrangements of experiments with Sherman and Austin correspond to an implementation of Skinner's terminology.

On the other hand, she has to acknowledge that "[c]ommunication as a process, is not dealt with in all detail in *Verbal Behavior*" (Savage-Rumbaugh 1984, 244), and "[u]nfortunately, Skinner does not go on to provide a vocabulary that does apply to the phenomenon of communication" (Savage-Rumbaugh 1984, 244). At the same time she quotes from Skinner, who brings in the concept of *intentionality* – namely *belief* – in order to conceptualize the behavior of a listener with respect to a speaker.

A departure from her attempts to conceptually unify intentionalistic and behavioristic approaches at the meta-level can be traced to her (1993) article, where she conceptualizes *reference* as that to which one points, to which one draws attention; "a specific thing, idea, goal or particular activity that is desired etc." (Savage-Rumbaugh 1993, 460), while the *act of reference* "is assumed to be an intentional act that is carried out for a specific purpose" (Savage-Rumbaugh 1993, 460) and which always takes place between individuals "sentient of the nature of the communicative system that they employ" (Savage-Rumbaugh 1993, 460).

This shift from a behavioristic to an intentionalistic conceptualization finds its continuation in (Rumbaugh and Savage-Rumbaugh and Washburn 1996). Here, for the first time, it is viewed as a turning away from a *natural-science-research paradigm*, because the behavioristic approach is now understood as an extension of concepts and research methods of physics and chemistry into the realm of life. This extension, however, brings in concepts and methods that are alien to life, while life is at the basis of the existence of animals and humans. The *methodological* consequences are then spelled out by Savage-Rumbaugh as a critique of the means available to *Science*, where Science is understood as the sum of accepted methods of her time in primate behavioral research (Savage-Rumbaugh 1999, 115). The critique runs as follows (Savage-Rumbaugh 1999, 161–162)

> The real difficulty here is that living organisms do normally interact, and the "observer" stance is not the same as the "participant" stance. We cannot treat primates like particles of matter, for which the mixing and treatment procedures of one chemist can be replicated by those of another. Primates have memory, and the history of one's past interactions determines the nature of future interactions. One participant is not equal to another because their histories are not equal. And an observer is not equal to a participant because an observer stands outside. ... In the arena of ape language, the participants are also the researchers. There are no other "observers" standing by observing the participants. ...

Observers, by themselves, cannot change Kanzi's perception of the woods, nor can Kanzi's behavior change their perceptions. Thus a true observer would never encounter the effect of language. By the current standards of Science, I must have an observer of all my actions and all Kanzi's actions in order to validate any report. ... Kanzi and I, however, are not like vials of sodium and chloride waiting to interact, unaffected by whatever means is used to join us or to watch us. We are constantly interacting, and the nature of our interaction is affected by the observer. Moreover, each observer affects the nature of that interaction differently. ... Thus Science, as currently structured, will not take the participant's account and does not recognize as real the effect of the observer.

The final meta-scientific and methodological shift in Savage-Rumbaugh's ALP came in the form of a turn towards *narrative ethnography*.[102] Its focus is the conduct of life of both and humans living together in a *Pan/Homo* culture, where through (Savage-Rumbaugh et al. 2005, 312–313)

> the living of a joined life, one learns about shared emotions, shared intentions, shared goals, shared perceptions of time, shared ethical norms, shared health and shared illness, and shared mythologies, among many other culturally instantiated ways of being. These shared perceptions of reality serve as a sort of clay from which the events of daily living become co-molded and co-interpreted. Experiential knowledge of these events becomes verifiable through simple daily acts of joint living and joint engagement. This happens as it does because if perceptions emerge that prove to be inaccurate, nearly all attempted joint actions will fail.

From the point of view of *methodology*, this argues for the scientific superiority of "participant-based ethnographic studies" (Savage-Rumbaugh et al. 2005, 311) that rely "upon insight, intuition, and analysis of the observers, who are ... participant observers in the classical anthropological tradition. Narrative accounts, by definition, describe events. They do not predicate events, nor do they focus upon quantitative data" (Savage-Rumbaugh and Fields 2006, 223).

At the same time, the turn of the ALP to narrative ethnography represents a *sociocultural perspective*; "We seek to innovate ... [by] emphasizing ethnographic facts of a Pan/Homo society that speak to the socio-historical heritage of Soviet psychology spearheaded by Vygotsky, Leont'ev, and Luria" (Fields and Sagerdahl and Savage-Rumbaugh 2007, 164).

---

102   On narrative ethnography see (Tedlock 1991) and (Tedlock 2004). "Some fieldworkers have used the term narrative ethnography to highlight *researchers'* narrative practices as they craft ethnographic accounts. This use features the vibrant interplay between the ethnographer's own subjectivity and the subjectivities of those whose lives and worlds are in view. These ethnographic texts ... take special notice of the researcher's own participation, perspective, voice, and especially of his or her experiences in relation to the experiences of those being studies" (Gubrium and Holstein 2008, 251).

Based on that turn and that perspective, she and her collaborators draw the conclusion that the ALP for a long time was based on at least four incorrect suppositions. First, it was supposed that the testing language of human experimenters was somehow *neutral* in respect to the subject-matter – "objects (= apes)" – to which it was applied in research. Second, it was not realized that the subject-matter to which it was applied was *also a language*. Third, it was not realized that Savage-Rumbaugh and, for example, Kanzi were communicating in the testing language, e.g., when the former explained to the latter the experimental arrangements, the language to be tested was much simpler than the testing language (Fields and Sagerdahl and Savage-Rumbaugh 2007, 182–183). This thus should witness "[t]he absurdity of ape language research" (Fields and Sagerdahl and Savage-Rumbaugh 2007, 182) in its *pre-ethnographic form*; the latter being characterized also *as science in its objectivistic claims* that Savage-Rumbaugh and her collaborators refuse to accept: "Our view is not the classic one of studying what is 'out' there according to one or another disciplinary perspective" (Savage-Rumbaugh *et al.* 2005, 312).

Fourth, (Savage-Rumbaugh *et al.* 2005, 312) the dominant view was that experiments with apes should yield predictably similar results, that is, they should lead to the discovery of *universals of behavior* independent of the rearing background of these apes. Contrary to this, once the shift to narrative ethnography with its sociocultural perspective took place, the ALP held to the view that experiments with subjects from different, particular *Pan/Homo* cultures will yield different results. With respect to Kanzi's stone tool-production, whose results lacked the features of stone tools produced in the Oldowan period, this means that (Savage-Rumbaugh and Fields 2006, 240)

> [f]rom a postmodernist's view, there is nothing universal about the cultures that produced Oldowan technologies ... The absence of Oldowan features in our stone tools is meaningless, unless one assumes that God is broadcasting Oldowan algorithms and you simply have to have the right kind of humanlike brain to access this universal.

6.5 Some Objections

With the foregoing as relevant background, let me now offer a *critique* of Savage-Rumbaugh's most recent methodological views. In her analysis of how to understand the communicative systems of primates she states that one faces a problem similar to the following (Savage-Rumbaugh *et al.* 1996, 173):

> what would happen if a scientist tried to decode an unknown human language by looking only at the relationship between the words of a speaker and the behavior of the listener. ... If an outside observer does not any understand any of the words, nor even

whether one is hearing individual words as opposed to phonemes, it is impossible to rely upon the correlations between the words of the speaker and the listener to gain an understanding of the meaning of the sounds emanating from the speaker.

An additional problem comes in due to the fact that even an (Savage-Rumbaugh et al. 1996, 173)

> [i]increasingly sophisticated grammatical analysis cannot provide greater insight into the basic phenomenon of symbolic communication, because the meaning transferred by the words often does not lie in grammar but in the mutually understood intent of the speaker.

A possible *way out* of these problems in the study and discovery of symbolic communication in non-human primates could be to approach it in the same way as we approach symbolic communication in our own species, namely, *via language*. This, then, means that the "primary problem, that of breaking into language," (Savage-Rumbaugh et al. 1996) can be accomplished only via *language*. Thus, one faces here a "chicken-or-egg" problem: *in order to trace and study language in a species—human and/or nonhuman—one already needs language.*

Accordingly, since symbolic communication is by its very nature permeated with *intents*, Savage-Rumbaugh emphasizes that our "current analytical and quantitative methods determining whether or not a purported 'fact' about animal communication is verifiable" do not allow us to break into that circle. Furthermore, we lack "the scientific tools to decode the communicative systems of other species in an objective manner" (Savage-Rumbaugh et al. 1996, 175). As a result, she proposes the following alternative method of approaching and identifying the language in a non-human species (Savage-Rumbaugh et al. 1996, 175):

1) The scientist asks: "If an ape ... were to have a language, what might it look like?"
2) The scientist then asks: "What might an ape ... need to do, that would require a symbolic system of communication?"
3) If the scientist succeeds in discerning a form of behavior in apes that requires language, then he/she could then investigate into their particular actions.

However, if one takes a closer look at the methodological norms proposed by Savage-Rumbaugh, one finds that she has not really proposed a way out of the circle of language. Even in step #1, above, she implicitly presupposes that anyone who tries to locate a language in the realm under investigation *already knows in advance at least what language is*. The same holds for the proposal listed under step #2. Only by knowing what language is can one know what types of behavior could involve language. Stated otherwise, *it is not possible, by using a language free of concepts referring to language – that is to say, from the position of an outside observer not employing language as a research "instrument" – to discern behavior in apes that requires language.*

This norm has to be stated even more robustly, in the case of attempts to track language in another species and establish intra-species communication: *One needs both the mastery of language and a meta-reflection on language in order to find out whether language is present in the "alien" species under investigation.*

As a test of my critique of Savage-Rumbaugh's proposals in the realm of methodology of the ALP, let us analyze her articles (Savage-Rumbaugh et al. 1996) and (Savage-Rumbaugh 1998), where she deals with the attempts to trace symbolic communication in feral bonobos. The issue at stake was to find out if separate subgroups of bonobos belonging to one group communicate when moving on the ground in a dense forest. The type of movement, the forest's presence, and the distance between the subgroups precludes both gestural and vocal symbolic communication. The task was to explain how it is possible that even in the absence of these two types of communication, still the subgroups were able to keep track of each other by some form of knowledge. The presence of the latter was evidenced by the fact that the subgroups always reunite at some distant feeding or nesting sites and usually take the same path to these sites. Now, the question for Savage-Rumbaugh was whether that knowledge was passed on from one subgroup to another by some form of communication. In order to answer this question, Savage-Rumbaugh drew on the knowledge given to trackers that guided her in the forest. The trackers were able to follow the routes of the subgroups by identifying vegetation altered by the bonobos in the following ways: (i) plants' leaves were tramped down by the bonobos' feet, (ii) branches were broken off and stuck perpendicularly into the ground, and (iii) plants were smashed in the middle of the path pointing in one direction (Savage-Rumbaugh 1998, 163).

Based on the identification of these alterations, Savage-Rumbaugh then stated the following methodological generalization (Savage-Rumbaugh et al. 1996, 179):

> To determine if symbols are being utilized, it is essential to begin by looking for some sort of telling pattern in the occurrence of events. If a pattern of similar symbolic events exists across many dissimilar instances, can a symbol system be assumed to be operative? This is true of human communication systems as well those of bonobos. For example, one of the communication systems used by the trackers was vegetation-based. ... This system was very inconspicuous, both by its nature and by the manner it was in which the trackers left the signals. The meanings of such symbols could not have been deciphered without the extensive effort had not the trackers explained it. A language-based explanation of a symbol defines for the listener the regularities of occurrence that are permitted for that symbol and thereby the events to which it can be linked and which

define its meaning. Lacking a bonobo translator, one can only work diligently to try to discover any patterns of regularity for oneself. ... Imagine trying to determine the pattern inherent in Morse code without knowing the language in which it was being sent.

But, this means that *only a being already using symbols for communication* (i) *can use symbols – in a meta-reflective move – to identify a symbolic pattern in a set of otherwise different events*; (ii) *knows what are the regularities of occurrence permitted for particular symbols, that is, he/she knows what are their respective meaning/intent and reference*. Finally, only by fulfilling conditions (i) and (ii) can the being (iii) *function as a translator*. In the example given above, the trackers functioned for Savage-Rumbaugh and her collaborators initially as a bridge into the (possible) communication between the bonobo subgroups, that is, as translators for feral bonobos.

Let us now turn to Savage-Rumbaugh's view that one lacks "the scientific tools to decode the communicative systems of other species in an objective manner" (Savage-Rumbaugh *et al.* 1996, 175), as well as her claim that "[o]ur view is not the classic one of studying what is 'out' there according to one or another disciplinary perspective" (Savage-Rumbaugh *et al.* 2005, 312).

If my critique above applies, then this means, first, that the attempts to find out if another species is (capable of) using a language have an *objective aim* – to find out a state of affairs in that species and what one can use in order to reach that aim is, and in fact can only be, language. This, in turn, means that in the process when a species' language should identify and recognize the existence and structure of another species' language, this identification and recognition not only yields claims with a pretension to *objectivity* and *truth* but also with an *essentialist* pretension.[103] The basic concepts and rules of the former language should reconstruct the rules at work in the latter language; *the former should lock on the latter by means of understanding*. Contrary to this, in the natural sciences (e.g., physics, chemistry) hypotheses are stated that initially do *not* have any objectivistic/essentialist pretension: the hypotheses with their concepts are introduced, initially, as pure *conventions* because they need not lock on to (understand) a pre-existing language system in the realm under investigation.

Second, the issue of objectivity and the need to apply natural and technical sciences to the realm of intra-species communication also surfaces when one looks at the history of the ALP from the early 1970's on. In order to find out if apes are capable at all of symbolic communication with humans and comprehension of spoken English, a great deal of effort was vested into finding ways to bridge the

---

103 Here I draw on (Habermas 2001, 9–10).

somatic differences between the apes and the human experimenters which distorted their (possible) mutual communication. This means that the claim that the ALP is part of the humanities, and that it proceeds by the methods of narrative ethnography, has as its background the (successful) attempts to restore the somatically distorted intra-species communication by the employment of technical means having their origin in the natural/technical sciences (e.g., the implementation of the lexigrams in the computers and their display on monitors).

The meta-scientific and methodological theories that put into opposition natural/technical sciences and humanities in the field of ape language research do not reflect the fact that all actual communication involves the employment of bodies.[104] This fact is usually hidden in the background in the case of *inter*-species communication, but surfaces when one deals practically/experimentally with attempts at *intra*-species communication.

Let us, finally, turn to the above mentioned denial of the existence of *transcultural universals*. The central question – with respect to that denial – for Savage-Rumbaugh and her collaborators is as follows: "Are there limitations to great ape cognition that are not culturally based? Is brain size and/or brain wiring a basic constraint in some yet unexplained way?" (Savage-Rumbaugh *et al.* 2005, 324). The answer they give runs as follows (Savage-Rumbaugh *et al.* 2005, 324):

> We know that human children who suffer hemispherectomies at an early age still acquire language and human culture ... even though their brains are half the normal size and clearly must become differently wired. These simple facts point us in a new direction of self-understanding and awareness. They emphasize the strength of cultural realities ... Minds do not arrive preformed and minds do not emerge in precise stages. Minds are bended and folded by culture forces that are operating at a level we are only beginning to understand.

But, this answer hopelessly contradicts the results of experiments accomplished in the framework of the ALP, namely, those comparing the cognitive/linguistic capabilities of Alia and Kanzi. The human infant took off in its development at a rapid space at the age between 2½ and 3 years, and clearly left behind the cognitive/linguistic capabilities of Kanzi (at that time already over 8 years old). *The elimination of differences in the sociocultural rearing conditions between Alia and Kanzi – on which those experiments were based – points to differences in universals given in the bonobo species and Homo species.* So, since these universals cannot belong to a single socio-cultural matrix, they can be explained only phylogenetically "as biological adaptation in the sense of an explanation by means of the

---

104   On this see (Krüger 1999) and (Krüger 2010).

theory of natural evolution" (Krüger 2010, 130), that is to say, *not by concepts of sociocultural disciplines but of natural science disciplines*. And these universals can be viewed as enabling structures, in the sense that they *enable* but *do not determine* such phenomena as intentionality and inter- and intra-species mutual understanding and communication.[105] This view, in turn, leads to a new direction in research that by its nature is of the *non-natural-science type*, namely, research into what those universals *enable* and what they *block off* in the sense of a comparison of intra-species *differences* in the early (infant) ontogeny of cognition, learning, tool-use, language, etc. This is the path taken by Michael Tomasello and his collaborators starting in the 1980's. A meta-scientific and methodological analysis of their findings is the subject-matter of the next chapter.

---

105   Here I draw on H.-P. Krüger's (2010) where the German term "Ermöglichkeitsstrukturen" is introduced for these structures.

Chapter 7: Varieties of Intentionality: Michael Tomasello

The aim of this chapter is to analyze Michael Tomasello's research into human infants, children, and great apes from the scientific, metascientific, and methodological points of view. I regard these three points of view as corresponding to the three dimensions of empirical sciences and at the same time making possible a deep analysis of such sciences. I shall utilize the first point of view to analyze Tomasello's introduction and employment of the central concepts of empirical research and theory construction in the framework of developmental and comparative psychology. In it I shall distinguish two its phases, one from the eighties and nineties of the 20th century, and another one from the new millennium and lasting till now. In the second point of view I shall deal with Tomasello's reflections on central concepts like first-order and second-order intentionality, shared intentionality, etc. Finally, from the third point of view I shall consider the methodological consequences of the choice and employment of concepts in the first and second dimensions, analyzing especially the methods employed by Tomasello in the practice of empirical research, in theory construction, and in explanation, that is, the relations of the *explanandum* to the *explanans*. Here I will draw partially on the works of Donald Davidson, Hans-Peter Krüger, and Jürgen Habermas.

7.1 The Scientific and Meta-Scientific Dimensions

As stated above, I shall provide a reconstruction of the practice and results of Tomasello's scientific research into great apes, human infants, and children involving test trials, empiric-quantitative procedures, and theoretical clarifications. In this practice there are two clearly distinct periods: one centered around the concepts of first-order and second-order intentionality and the other centered around the concept of shared intentionality.

7.1.1 First order and second order intentionality: the eighties and nineties

In the first period of his endeavor in the field of developmental and comparative psychology, Tomasello put the emphasis on what unifies human infants with great apes, noting that both belong to the order of primates (*primata*), and what makes them different. What makes them similar is that they exchange with their conspecifics *signals* which they do not send intentionally, that is, without any intent on the part of the sender, while communication still takes place between the sender and the receiver because the latter is capable of interpreting the signals. The acts of sending of signals are labeled *perlocutionary acts* and are different from *illocutionary acts*, which stand for acts intentionally oriented toward the receiver; that is, acts in which an alternation of the attention of the sender takes place between an object and the conspecific/receiver with whom the sender wants to communicate about the object.

In one of his first articles from this period (Tomasello *et al.* 1985) he subjected to experimental trials infant and juvenile chimpanzees in order to find out if, and if so, how, they acquire and use intentionally produced gestures in the instances like obtaining food from adults (nurse, food-beg), soliciting specific adult behavior (groom, tickle, ride/walk with), and engaging in certain types of play (chase, rough-and-tumble play). The behavioral categories and some of their corresponding gestures are related in the following table (Tomasello *et al.* 1985, 178).

Table 7.1: Behavioral categories in chimpanzees and some of their corresponding gestures

Behavioral category	Corresponding gesture
Nurse	Infant touches the side of the mother to get her attention when she is not looking
Walk/ride	Infant approaches the adult with arms extended and places them on his/her back
Groom	Infant puts the hand of the adult on the spot to be groomed
Tickle	Infant puts the hand of the adult under arms
Food from adult	Infant sucks the lower lip of the adult or places the hand under adult's mouth
Play	Infant raises arm and charges other or slaps ground and looks to other

The gestures listed here were viewed by Tomasello as *ritualizations*, that is, as incipient movements of social behavior. As to the process of acquisition of these gestures, he viewed as central the process of individual creativity in which the chimpanzee produces "just the first part of behavior for the purpose of obtaining a result originally produced by means of a longer behavioral sequence" (Tomasello et al. 1985, 184).

The issue of learning was then extended by Tomasello to the cases of the learning of tool use and of gestural communication in chimpanzees over a prolonged period of time (Tomasello and Gust and Frost 1989). In the first extension, the conceptual options were as follows: learning the use of tools by way of the individual's own *discovery*, or, by *observation* of adults' tool use, or, finally, via precise *imitation* of adults' tool use behavior. In order to find out which of these options was at work, a testing procedure was designed and then implemented. In it, an adult chimpanzee acting as an instructor demonstrated the use of a T-shaped metal bar to pull food into the cage where she was placed. The chimpanzees subjected to the trials were split up into two groups: the experimental group and the control group, the latter being subject to a control condition, namely, that chimpanzees from it were not given the instruction of the T-bar use by the adult chimpanzee.

The behavior of the chimpanzees was classified into five groups: inactive, attempt to obtain food without the use of the tool, subject plays with the tool (touches/grasps it), attempts to use the tool to obtain the food, and, finally, successful attempts. The results of test trials were as follows. At the beginning, none of the subjects in either group was able, in ten pretest trials, to use the tool. Eventually, in later trials, in the experimental group all four subjects learned the use of the tool. In the control group, however, two of the three subjects were completely unsuccessful, while the third one displayed only inconsistent success. Finally, none of the subjects in either of the groups was able to imitatively copy the procedure of the adult demonstrator.

From these experimental results Tomasello drew the conclusion that the young chimpanzees learned via observation that the T-bar can be used as a tool to obtain food. At the same time, however, he did not view their learning of the tool use as a case of *imitative learning* because none of the chimpanzees in either group copied in a precise and detailed way the specific strategy of employment of the T-bar. Thus, in the eighties and nineties, Tomasello viewed *emulation learning*, that is, learning of tool use by trial and error performed by an individual, as the dominant type of learning in young chimpanzees for which they display a strong capacity.

This interpretation is based on a threefold conceptual differentiation of the types of tool use. The first stands for the case when the aspect of the *very tool* becomes salient ("enhanced") after the demonstration; the second stands for the case when the goal is reproduced together with the tool use in an unspecified manner. Finally, the third stands for the case when the imitator tries to copy *precisely the actual behavior* of the demonstrator.

A concept of an additional type of learning symptomatic for apes was introduced in (Tomasello 1990), namely, that of *conventionalization*. It stands for individual learning processes taking place in particular social interactions and is used to acquire/learn communicative signals from conspecifics. Here, according to Tomasello, social behavior is transformed into signals produced *intentionally* by the sender, but no signal by itself is transferred from the sending to the receiving organism; thus, it is not a semiotic relation (Tomasello 1990, 295). Such a relation takes place only if the sender uses an abbreviated behavioral sequence that stands for a behavioral type. If one understands under *"social learning"* individual learning influenced by the social environment and which in the case of a more active role of such environment can be labeled "local (or stimulus) enhancement," under *"cultural learning"* a learning based on the understanding of other persons' beliefs and intentions (Tomasello and Kruger and Ratner 1993, 495-496), and under *"instructed learning"* learning, in which teaching is involved in the sense that that "learners interiorize the instructions of teachers and use them consequently" (Tomasello and Kruger and Ratner 1993, 497), then the following table characterizes learning processes in chimpanzees (Tomasello 1990, 303):

Table 7.2: Learning processes in chimpanzees' tool use and communication

Type of Employed learning in	Individual	Social	Cultural
Tool use	Strong evidence for environmental shaping	Strong evidence for stimulus enhancement and emulation	Weak evidence for imitation and no evidence for instructed learning
Communication	Strong evidence for conventionalization	_	Weak evidence for imitation and no evidence for instructed learning

Compared to the above, the types of social learning and their features in human infants can represented via the following table (simplified from Tomasello and Kruger and Ratner 1993, 503):

Table 7.3: Types of social learning in human infants

Characterized by	Types of cultural learning processes	Imitative (age 9-12 months)	Instructed (age 3-4 years)	Collaborative (age 6-7 years)
Social-cognitive ability		Perspective-taking (e.g., imitative learning)	Intersubjectivity (e.g., intentional deception)	Recursive intersubjectivity
Concept of person		Intentional agent	Mental agent	Reflective agent

Based on such a typology of learning processes, Tomasello tries to solve the problem of existence of *chimpanzee culture(s)*, and at the same time to differentiate between a (possible) chimpanzee and human cultures. Since there is no evidence for cultural transmission in chimpanzees in the fields of tool use and communication, Tomasello differentiates conceptually (Tomasello and Kruger and Ratner 1993, 495) between chimpanzees' *social groups* and humans' *cultural groups*. The latter he characterizes via their products, namely, material artifacts, social institutions, behavioral traditions, and languages. What is symptomatic for human cultures is that (Tomasello and Kruger and Ratner 1993, 495)

> they accumulate modifications over time ... This accumulation of modifications over time is often called the "ratchet effect", because each modification stays firmly in place in the group until further modifications are made. No cultural products exhibiting anything like the ratchet effect have ever been observed in the ontogenetically acquired behaviors or products of nonhuman animals.

In order to make the conceptual differentiation between emulation learning (learning of tool use by trial and error performed by an individual), local enhancement (learning by being attracted to a locality), and ritualization – all belonging to the group of social learning, on the one hand, and cultural learning, on the other, Tomasello brings in the concept of *intentionality*, so that *cultural learning by an individual relies crucially on the understanding of intentions of the individual's conspecifics* (Tomasello and Call 1994).

If one understands the term "intentionality" as "to relate contentfully to objects and states of affairs, to represent them" (Rakoczy 2008, 2) and under "intentional" one understands "acting with an intention" (Tomasello and Rakoczy 2003, 122), then Tomasello's conclusion from the nineteen eighties and nineties was that while a certain type of intentionality is common to human infants and chimpanzees, there exists a type of intentionality which is symptomatic exclusively for human infants. The former is the so-called *"first-order intentionality,"* standing for the ability to relate to entities in the external world via perception and memory, as well as for the ability to choose the "goals, behavioral means for pursuing goals, attentional foci for monitoring progress toward goals" (Tomasello and Call 1997, 405). The latter is *"second-order intentionality,"* standing for the knowledge of an individual about the intentional states (e.g., perceptions, goals, beliefs, etc.) of other subjects, that is, "an understanding of how organisms' intentional interactions with the environment are organized" (Tomasello and Call 1997, 405).

This means that in the comparison of great apes' behavior with that of human infants one has to discern the subjects' behavior to (inanimate) objects from their behavior to subjects, both conspecific and from other species. In respect to the former, Tomasello's experimental trials showed that great apes are capable of understanding causality that is at work in objects, but they fail to understand "social causality" that is at work in interactions between subjects (Visalberghi and Tomasello 1998).[106]

Based on Tomasello's view on the difference – at the level of the manifest types of intentionality – between great apes and human infants – the *explanatory strategy* used by him in the eighties and nineties can be represented as follows:

Figure 7.1: Tomasello's explanatory strategy in the eighties and nineties

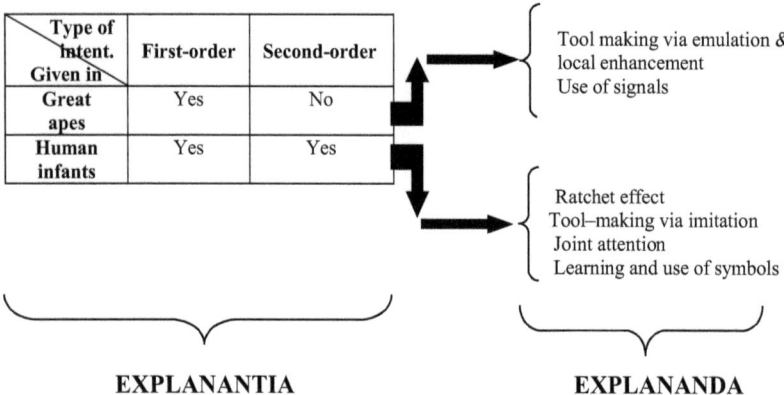

In the second period of his scientific endeavor, Tomasello brings in not only important changes to the explanantia but turns the latters into explananda.

7.1.2 From first order and second order intentionality toward shared intentionality: the new millennium

The first shift in the views of Tomasello in the new millennium can be traced back to his article (Tomasello 1999b). Here he starts to reflect on the basis of the differences between chimpanzees and human infants as represented above in Figure 7.1. This basis should be, he claims, a *biological* adaptation for culture absent in other primates, and here "the key adaptation is one that enables individuals to understand other individuals as intentional agents like the self" (Tomasello

---

106 On the use of symbols and signals in the context of this type of "causality" see (Tomasello 1996).

1999b, 509). The biological adaptation by which the species *Homo sapiens* differs from that of *Pan troglodytes*, has its origin in the genome of *Homo sapiens* that, however, differs from that of *Pan troglodytes* only by 1% to 2%. Tomasello also relates that biological adaptation to the types of intentionality manifest in human infants and chimpanzees, as well as these types of intentionality to phenomena like imitative learning, ratchet effect,[107] perceptual and sensory-motor representations, emulation learning, and ontogenetic ritualization. His view on the relation between these concepts as given in (Tomasello 1999b) can be represented by the following figure:

Figure 7.2: Tomasello's explanatory strategy in the new millennium

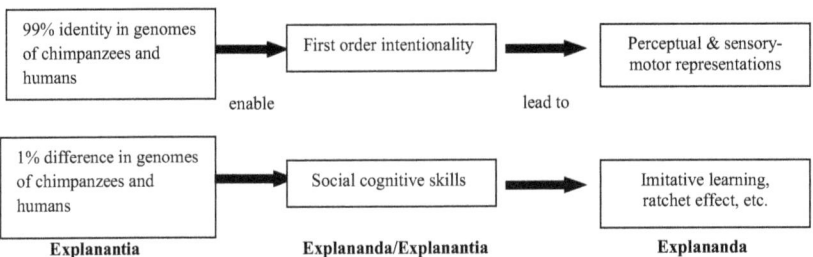

Drawing on the concept of biological adaptation, Tomasello draws the conclusion that a genetic event led to *Homo sapiens*' specific social cognition and to the social-cultural transmission process characteristic of it. This then (Tomasello 1999b, 526)

> led to a series of cascading sociological and psychological events in historical time ... And so, while not denying that a significant genetic event happened in human cognitive evolution, I would deny that this event specified the detailed outcomes we see in adult humans today. In my view, that genetic event merely opened the way for some new social and cultural processes that then, with no further genetic events, created many, if not all, of the most interesting and distinctive characteristics of human cognition.

That genetic event is a specific type of biological adaptation, namely, for understanding other beings as intentional agents. This type of adaptation is a necessary but not a sufficient condition for the actualization of that understanding, since the latter is a result of the process of ontogeny to which correspond no specific biological underpinnings. As emphasized by Tomasello in respect to human cognition (Tomasello 1999a, 11),

---

107 On the ratchet effect see (Tennie and Call and Tomasello 2009).

historical and ontogenetic processes ... are enabled but not in any way determined by human beings' biological adaptation for a special form of social cognition. Indeed, my central argument ... is that it is these processes, not any specialized biological adaptations directly, that have done the actual work in creating many, if not all, of the most distinctive and important cognitive products and processes of the species *Homo sapiens*.

That concept of *enabling* of cultural cognitive skills in humans stands for Tomasello's attempt at an adequate unification of "both cultural and biological dimensions of human cognitive development" (Tomasello 2000, 37), and where such an unification has to be accomplished via the conceptual unification of "(a) an evolutionary approach to the human capacity for culture, and (b) an ontogenetic approach to human cognitive development in the context of culture" (Tomasello 2000, 37).

Another important shift in Tomasello's views in the new millennium, as compared with his earlier views, is that in a battery of tests he has shown that great apes display *not only first order intentionality but also second order intentionality*. That is, they *know what others see and do not see*, and *what others know and do not know*, and thus, they *know the mental states of others*.[108]

As an example, let me deal with the paper (Call 2001) which describes the attempt to find out what a chimpanzee knows about the perceptual knowledge of his/her conspecifics. The basis of the test trials was the competitive situation obtained in pairs of subordinate-dominant chimpanzees competing for two pieces of food located in a cage. The chimpanzees were placed into two cages on opposite sides of a third one, so that initially, only one of the two pieces of food in the third cage was visible to the dominant chimpanzee, while the second piece was not visible to him due to the placement of an opaque barrier. On the other hand, both food pieces were visible to the subordinate chimpanzee. The subordinate chimpanzee could make his choice before he saw to which piece of food the dominant chimpanzee would move. The result of experiments of this type was that the subordinate chimpanzee picked up significantly more of the pieces of food which were not visible to the dominant ape than those which were visible. In a control experiment, the opaque barrier was replaced by a transparent one, and its result was that the subordinate ape's preference for the hidden food disappeared. The general conclusion from these types of experiments was that the subordinate ape knew what the dominant ape could and could not see.[109]

---

108   On this see (Hare and Call and Tomasello 2000), (Tomasello and Call 2006), (Schmelz and Call and Tomasello 2011) and (Tomasello and Call and Hare 2003).
109   (Call 2001) describes also other types of tests; their results were the same.

In addition to conceptually broadening the capacity of great apes to second order intentionality, Tomasello also brought in the second millennium the concept of *shared intentionality*, which would enable him conceptually to differentiate between intentionality displayed by humans from that displayed by great apes.[110] Their difference is based not on "individual inventions arising out of human's extraordinary brain power, but rather ... [on] collective cultural products created ... over historical time" (Tomasello and Rakoczy 2003, 121). He thus views linguistic communication, shared pretense, and discourse about mental processes as the uniquely human forms of social and collective intentionality. These forms he characterizes as *ontogenetic constructions*, and makes the following three basic claims about them (Tomasello and Rakoczy 2003, 122–123):

- Human beings have a biological adaptation for a species-unique form of social cognition. This adaptation expresses itself ontogenetically as two key developmental moments, one at about one year of age and one at about four years of age ... these are ... two phases of the same developmental pathway: understanding persons as intentional agents and then as mental agents ...
- Understanding and coordinating with intentional agents at one year of age ... enables human children to participate in and master cultural activities of all kinds, including linguistic communication ...
- Three and four year old children's coming to understand mental agents ... depends both on the understanding of intentional agents and on several year period of continuous interaction ... with other persons. Based especially on their participation in perspective-shifting and reflective discourse, some new kinds of normativity emerge—specifically, those involving beliefs ..., which in turn enable the comprehension of cultural institutions based on collective beliefs and practices ... We thus call these activities collective intentionality.

The various "faces" of intentionality are further conceptually explicated in (Tomasello *et al.* 2005). According to Tomasello, "human beings, and only human beings, are biologically adapted for participating in collaborative activities involving shared goals and socially coordinated plans of actions" (Tomasello *et al.* 2005, 676). Tomasello subsumes mutually shared goals and socially coordinated plans of actions under the term *"joint intention."* This type of intention is

---

110   Tomasello viewed in 2008 the claim that shared intentionality is *Homo sapiens*' *differentia specifica* when compared with great apes as a hypothesis; "the jury is still out" [Michael Tomasello's interview on 3SAT television station from October 30, 2008, 1.19 minute]. The same claim is made in (Call 2009).

neither of the first nor of the second type of intentionality, but of the *shared type of intentionality* or "*we-intentionality*."[111]

In order to give a precise understanding of the shared (or we-) intentionality, Tomasello deals anew with the conceptual triad: *intentional action – understanding intentional action – shared intentionality*. "Intentional action" is understood, drawing on the theory of cybernetics, as displaying the following three central features: action has a mental/thought goal; this goal should be attained in the environment; the actor perceives the environment and monitors the results attained in the environment.

In the case of one-year old human infants' understanding of intentional action given, he distinguishes three levels: (a) animate acting; "An observer perceives that the actor has generated his motion autonomously; that is, she distinguishes animate *self-produced* action from inanimate, caused motion" (Tomasello et al. 2005, 678); (b) the pursuit of goals; "An observer perceives and understands that the actor has a goal and behaves with persistence until reality matches the goal" (Tomasello et al. 2005, 678); an (c) the choice of plans; "An observer perceives and understands that the actor considers action plans and *chooses* which of them to enact in intentional action" (Tomasello et al. 2005, 678).

Finally, for *shared intentionality* holds that "the goals and intentions of each interactant must include as content some of the goals and intentions of the other" (Tomasello et al. 2005, 680); hence, it has the character of "collaborative interactions in which participants have a shared goal (shared commitment) and coordinated action roles for pursuing that shared goal" (Tomasello et al. 2005, 680).

Based on the above-mentioned experiments with great apes, Tomasello emphasizes that they understand the intentional actions of their conspecifics in terms of their perception, knowledge, and goals, but, unlike their human counterparts, they are unable to engage in higher order intentional actions, like instructing their conspecifics or engaging in cultural learning. This difference between humans and great apes he views as rooted in a *gap* (Tomasello and Moll 2010) in the realm of their respective cognitive skills, namely, those skills related to the *social realm*,[112] despite the fact that their skills display a high degree of similarity when dealing with the "*physical*" world; its causal relations, space determinations, and quantitative characteristics.[113]

---

111 On shared intentionality see also (Tomasello and Carpenter 2007) and (Pika and Zuberbühler 2008).
112 For a comparison of these skills see also (Call 2009).
113 On great apes' cognition in the sphere of "protophysics" see, e.g., (Albiach-Serrano and Call and Barth 2010); on great apes' cognition in the sphere of "protomathematics"

7.1.3 Shared Intentionality: from individual intentionality via joint intentionality to collective intentionality

The meaning of the term "shared intentionality" underwent in recent years a further specification in (Tomasello 2012), (Tomasello *et al.* 2012) and (Tomasello 2014) which was explicitly unified with a specific set of methods.

Due to that specification the meaning of that term should now be understood as "collaborative phenomena in general" (Tomasello *et al.* 2012, 674) and under which should be subsumed two distinct types of intentionality labeled as "joint intentionality" and as "collective intentionality." The former should stand for "for the ad hoc temporary collaboration characteristic" while the latter "for the more impersonal yet permanent group-minded practices and modes of collaboration that characterize cultural groups as a whole" (Tomasello *et al.* 2012, 674).

In addition to these two types of intentionality, he introduces yet another type he labels as "individual intentionality." These three types are embedded by him into a hypothesis about the evolution leading to the cooperative characteristics of the species Homo *sapiens sapiens*. This hypothesis, labeled as *interdependence hypothesis*, should explain how "at some point humans created lifeways in which collaborating with others was necessary for survival and procreation ... interdependent collaboration ... helps to explain humans' unique forms of cognition and social organization" (Tomasello *et al.* 2012, 674).

According to this hypothesis one can distinguish three phases of that evolution.

Phase 1 is characterized by a type of organism which is cognitively competent in the sense that it (Tomasello 2014, 8)

> operates as a control system with reference values or goals, capacities for attending to situations causally or intentionally "relevant" to these reference values or goals, and capacities for choosing actions that led to the fulfillment of these reference values or goals (given the causal and/or intentional structure of the situation).

One thus faces here a (Tomasello 2014, 9)

> flexible, individually self-regulated, cognitive way of doing things ... Within this self-regulation model of individual intentionality, we may then say that thinking occurs when an organism attempts ... to solve a problem, and so to meet its goal not by behaving overtly but, rather, by imagining what would happen if it tried different actions in a situation ...before actually acting. This imagining is nothing more or less than the "off-line" simulation of potential perceptual experiences.

---

see (Beran 2004), (Hanus and Call 2004), (Hanus and Call 2007), (Hanus and Call 2008), (Hanus and Call 2009) and (Rumbaugh and Savage-Rumbaugh and Hegel 1988).

Phase 2 is characterized by "the creation of a novel type of small-scale collaboration" (Tomasello 2014, 5), whose participants create goals they mutually share as well as joint attention to certain entities in the world they mutually share in an *ad hoc* manner. This leads to the creation of both individual roles and perspectives in this world. And in order to coordinate these roles, a new way of communication among these individuals by pointing and pantomiming arises. Here should originate *joint intentionality*, understood as "collaboration and communication between ad hoc pairs of individuals" (Tomasello 2014, 5).

Phase 3 should correspond to modern humans characterized by group mindedness among members of a cultural group, by the creation of a common cultural ground by means of shared cultural conventions, norms, institutions and mutual communication by a conventionalized language (Tomasello 2014, 5).

Tomasello, then, provides a justification of this three-phase hypothesis by a set of methods with which I will deal in part 7.2.2.

7.2 The Methodological Dimension: Explanantia and Explananda

In Figures 7.1 and 7.2 I used the terms "explanantia" and "explananda," but without reconstructing the ways Tomasello accomplishes the movements from the former to the latter. These movements are related to a whole set of methodological problems that were delineated in the last decade or so by D. Davidson, J. Habermas and H.-P. Krüger, and which I shall try to solve in this part of the chapter.

7.2.1 Davidson, Krüger, Habermas, and the explanantia/explananda in the developmental and comparative psychology

Davidson addressed in his article (1999) the problems centered around the genesis of various forms of intentionality as an explanatory problem, even if he did not use explicitly the terms "explanandum" and "explanans." He declares (Davidson 1999, 11):

> The difficulty in describing the emergence of mental phenomena is a conceptual problem: it is the problem of describing the early stages in the maturing of reason, the stages that precede the situation in which concepts like intention ... have clear application. Both in the evolution of thought in the history of mankind, and in the evolution of thought in the individual, there is a stage at which there is no thought followed after a lapse of time by a subsequent stage at which there is thought. To describe the emergence of thought would be to describe the process which leads from the first to the second of these stages. What we lack is a satisfactory vocabulary for describing the intermediate steps. ... We have many vocabularies for describing nature when we regard it as mindless, and we

have a mentalistic vocabulary for describing thought and intentional action; what we lack is a way of describing what is in between. ... It is not that we have a clear idea what sort of language we could use to describe half-formed minds; there may be a very deep conceptual difficulty or impossibility involved. That means there is a perhaps insuperable problem in giving a full description of the emergence of thought. I am thankful that I am not in the field of developmental psychology!

Krüger explicitly does use the terms "explanandum" and "explanans, and locates them in Tomasello's double distinction of possible understanding:

(i) understanding of actors as animate beings versus understanding of objects as inanimate entities;
(ii) understanding of actors as intentional agents versus understanding of actors as mental agents.

This double differentiation plays a twofold role in the works of Tomasello (Krüger 2010, 130–131):

It is, on the one hand, valid as a biological adaptation in the sense of an explanation via a theory of natural evolution. Then it had to be explained as a result of mutually independent processes of variation and selection. From this point of view it would come to the fore as that that is in need of explanation, that is, as the *explanandum*. The double differentiation, on the other hand, functions as an enabling structure for social-cultural performances ... The mentioned biological adaptation does not *determine* but *enables* ... As long as the biological adaptation turns into an enabling structure in understanding, does it apparently play the role of the *explanans*. The understanding of conspecifics as intentional and mental actors appears then as an *explanans* that explains the facticity of social-cultural performances.

So, the explanation in the framework of Tomasello's project in the new millennium would be complete if one succeeds (Krüger 2010, 164):

(a) to explain phylogenetically the forms of shared intentionality and an elementary language from a biological adaptation, and (b) in turn use this explained as an *explanans*, in order to explain the ontogenetic contribution in the accumulation of culture via the ratchet effect.

Finally, Habermas deals with the problem of explanans/explanandum from the point of view of detranscendentalization of the conditions of any and all possible cognition (*Erkenntnis*). These conditions should be, according to Kant, atemporal, but, as we know nowadays, they have – both at the level of ontogeny and phylogeny – a beginning in time. Due to this detranscendentalization, the very concept of the transcendental undergoes a profound transformation; "The transcendental rules are not ... outside the world ... and have a beginning in time. And ... the very transcendental conditions of the epistemic access to the world must be understood as something in the world" (Habermas 1999, 27; 2005a, 17). What is at stake here is (Habermas 1999a, 29–30; 2005a, 20–21)

> the peculiar status of rules that should retain the power of spontaneous generation of the world even though no one can claim any more that they are without origin. They should enable us to access the world, yet should themselves not drop down from the heavens, but take a status in the world. Such a type of entities obviously eludes the transcendental difference between world and innerworldly, to which corresponds the methodical dualism of understanding and observation.

One should make an attempt to unify the "hermeneutic approach to the structures of mind (*Geist*) embodied in the lifeworld with the biological explanation of their genesis" (Habermas 1999, 30; 2005a, 21). However, according to Habermas, not only do all the attempts at a unification of these complementary ways of consideration "lead to competing descriptions that cannot be seamlessly translated into one another" (Habermas 1999, 30; 2005a, 21), but the task to explain the intersubjective enabling conditions of experience in general lead to a paradox; "An explanans that explains the origin of the transcendental conditions would already itself have to obey the conditions named in the explanandum" (Habermas 1999, 31; 2005a, 21)

7.2.2 The structural, the structural-genetical, the structural-historical, and the historical-genetic methods in Tomasello's explanation of cooperative communication

A) The Structural Method

The first and most simple method employed by Tomasello *et alia* pertains to thought operations and the experimental trials based on them, whose aim is to grasp conceptually the structure of certain processes with an emphasis on the coexistence of the features and characteristics of this structure while temporally abstracting from their diachronicity.

Let me examine the paper (Hanus *et al.* 2011) in order to reconstruct the main features of this method. The aim of this paper was to test if great apes – chimpanzees, gorillas and orangutans – both from a zoo and from two sanctuaries, are capable of insightful behavior in a goal-directed action, namely, in the retrieval of a peanut out of a tube by spitting water into it, so that the peanut starts floating close to the top and can be retrieved by the ape. The central aspect of the experiments, from the point of view of methods, was a variation of several conditions under which the experiments took place. Finally, the results of the experiments with great apes were compared with the results of experiments with human children at the age of four, six, and eight years.

In the first experiment, the behavior of 19 chimpanzees and 5 western lowland gorillas (all from a zoo in Leipzig, Germany) was tested in the task of retrieving the peanut of a Plexiglas tube, closed at its bottom and open at its top, by collecting water from a dispenser, placed at the experimental site long before the tests started, and then spitting it into the tube that was filled already with water by ¼. The result of the experiment was that none of the chimpanzees and gorillas retrieved the peanut, none of them added water into the tube, and none of them found the appropriate solution to the task they faced.

In the second experiment, conditions of the experiment were varied among the group of apes, now consisting of 10 orangutans from a sanctuary in Indonesia and 25 chimpanzees from a sanctuary in Uganda. In both, new water dispensers were installed just few days before the experiment started. Another variation pertained to the presence/absence of water in the Plexiglas tube, so that the experiment was performed under the so-called *wet condition* versus the so-called *dry condition*; one half of the apes were tested under the former and the other half under the latter conditions. If an ape in Uganda from the wet group failed in the first four trials, in the fifth trial the amount of water was doubled, and if an ape from the dry group failed in the first four trials, in the fifth trial water was added to the tube, and if failure continued, water was added again. Here the guiding idea of the variation of the test conditions was to find out if additional information – the addition of water – would improve the inventiveness of the apes.

The tests described so far pertained to the *experimental* phase of the experiment. In addition, chimpanzees who were successful in this phase were subject to tests under three conditions in the *control* phase: *top* condition – the peanut was glued to the top of the tube and so the ape could retrieve the peanut; *table* condition – the peanut was placed on a table in front of the tube but was beyond the reach for the ape; *dry* condition – as before, the peanut was at the bottom of a dry tube – only here did spitting represent the adequate method for obtaining the peanut.

The result in the experimental phase was such that none of the orangutans (10 in total) from the dry and wet groups solved the task; from the chimpanzees (24 in total) two from the dry group and three from the wet group solved the task. In the control phase, water was added more often in the dry conditions than in the wet conditions, while the number of spitting did not vary between the table and the top condition. Thus, the conclusion, based on such a variation in the control phase, was that the chimpanzees added water only when it was really needed in order to retrieve the peanut. Based on the mentioned variations, the general conclusion drawn from the results of the tests was that the presence of water – as an additional piece of information – did stimulate the apes to reach the goal. The difference between the apes behavior in the case when the peanut was in the tube and the behavior when the peanut was outside, showed that they added water only in the former case. Thus, the apes displayed not only a goal-directed but also an insightful behavior. In the last experiment, 72 children were subject to an experiment with a pitcher, instead of a dispenser, as the water source.

B) *The Structural-Genetical Method*

In addition to the type of thought and experimental procedures exemplified in (Hanus *et al.* 2011), one can discern also another type of procedure related to the ontogeny of the cognitive and practical skills of human infants and children. It is based on what Tomasello labels as the cooperative communication, namely, that humans reason cooperatively; (Tomasello 2008, 94):

> humans [signal] their communicative intentions ... when humans see that someone is attempting to communicate with them, they want to know what he is attempting to communicate at least partially *because* he wants them to ... and they choose a response ... at least partially *because* that is what the other wants them to do.

This view of human cooperative communication is expressed as a model in the following figure relating the different components of that model and their relations ("C" stands for a communicator and "R" for a recipient) (Tomasello 2008, 98):

Figure 7.3: Cooperative model of human communication

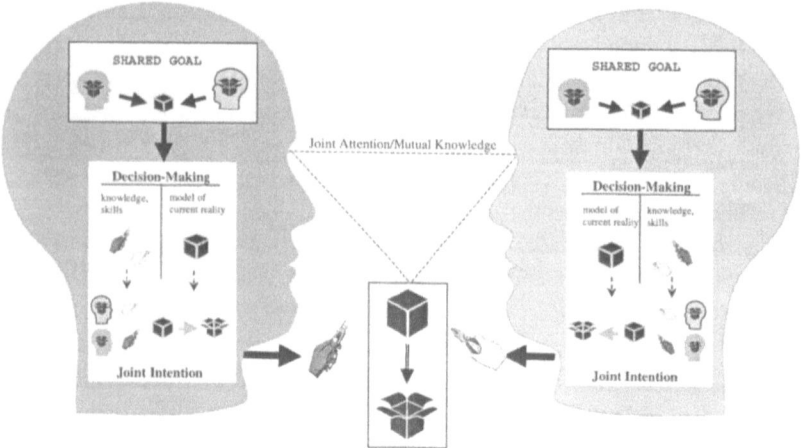

The construction of the cooperative model of human communication is also based on a comparison of human communication with apes' signaling. According to Tomasello, (2008, 93–94):

when apes observe another ape signaling to them, they try to discern what he wants via individual practical reasoning about his goals and perception. But they are not trying to understand the message *because* he wants them to, since the two of them do not share an assumption that he is trying to be helpful. The communicator thus does not signal or "advertise" his intention specifically, as humans do in signaling their communicative intention.

A special feature of joint attention and of the creation of a common ground in cooperative communication between two persons is recursivity, that is, "each of them sees, knows, or attends to things that she knows the other sees, knows, or attends as well—and knows that the other knows this about her as well, and so recursively potentially ad infinitum" (Tomasello 2008, 94–95).

Based on these concepts, Tomasello's view on the infrastructure of human cooperative communication, drawing on an understanding of apes' signaling, can be expressed as follows (Tomasello 2008, 105):

Table 7.4: *Infrastructure of apes' signaling and of human infants' and humans' cooperative communication*

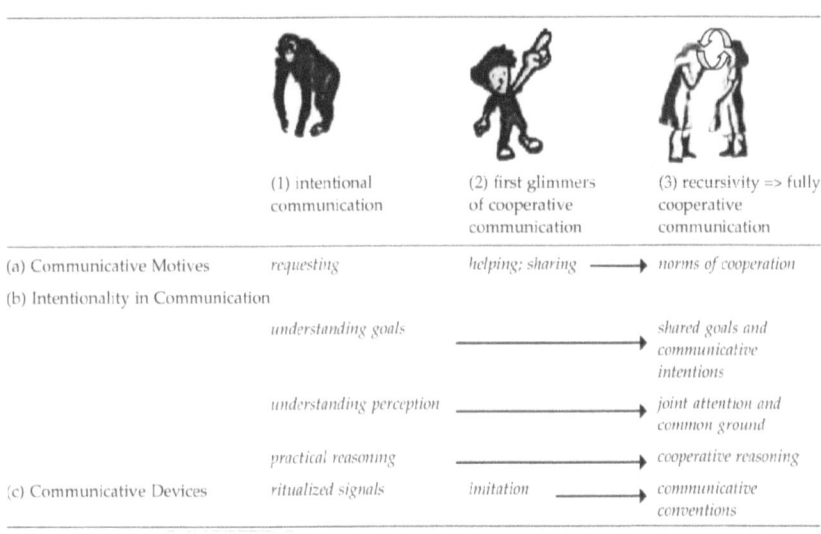

	(1) intentional communication	(2) first glimmers of cooperative communication	(3) recursivity => fully cooperative communication
(a) Communicative Motives	*requesting*	*helping; sharing* ⟶	*norms of cooperation*
(b) Intentionality in Communication			
	understanding goals	⟶	*shared goals and communicative intentions*
	understanding perception	⟶	*joint attention and common ground*
	practical reasoning	⟶	*cooperative reasoning*
(c) Communicative Devices	*ritualized signals*	*imitation* ⟶	*communicative conventions*

The flow diagram as given in Figure 7.3 and pertaining to a sequence starting from individual goals and leading to referential intention on the side of the communicator, and from reference to action on the side of the recipient, leads to the following question. What is the sequence of the components of human cooperative communication from the point of view of ontogeny, that is, how human cooperative communication works ontogenetically (Tomasello 2008, 109)? According to Tomasello, the ontogenetic sequence should be as follows:

Figure. 7.4: Ontogeny of human cooperative communication

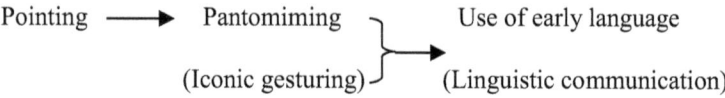

Here I spell out only the main features of infant pointing.[114] At its basis are, according to Tomasello, three general classes of social intentions/motives: (a) sharing of emotions and attitudes with others; (b) informing others of useful/interesting things; and (c) requesting others for help in order to attain goals.

From those three classes, sharing and requesting appear in the first months of infant's life and are, contrary to the informing motive, tied neither to an understanding of others' goals nor to an understanding of knowledge/information; that is, these two motives appear before shared intentionality is in place (Tomasello 2008, 135–141). These two classes can be viewed as the basis of intentional communication, and only when they are unified with shared intentionality, does this communication turn into *cooperative* communication. Here, however, Tomasello faced the following problem. Even if he claimed that *shared* intentionality is at the basis of infant pointing in cooperative communication, still it appears in infant's ontogeny simultaneously with his/her *individual* intentionality. In order to discern which of these two types of intentionality is the essential element for cooperative communication, Tomasello brought in data comparing typically developing human infants with chimpanzees and children with autism. What came out of such a comparison is the following. Great apes are intentional agents and they understand the intentionality of their conspecifics and of humans, but they do not participate in collaborative activities, that is, in activities "with an intentional structure comprising both a joint goal and complementary roles" (Tomasello 2008, 176). For example, in the context of food acquisition they are not other-regarding, are indifferent to inequity in this context, and focus only on their own gains and losses (Jensen *et al.* 2006). Children with autism – a

---

114   For a detailed analysis see (Liszkowski *et al.* 2004; 2012), (Liszkowski 2004; 2005), (Liszkowski *et al.* 2006; 2007), (Liszkowski and Carpenter and Tomasello 2007a; 2007b), (Tomasello and Carpenter and Liszkowski 2007).

neurobiological disorder – display a whole set of impairments: a deficit in imitation and in the use of joint attentional behavior, a reduced frequency in responding to others' bids for joint attention, and an absence of gesturing to share interest in an object.[115]

Based on these data, Tomasello unified the descriptions of typically developing infants, autistic children, and great apes, and then provided the data for the emergence of the four main social-cognitive skills in them (Tomasello and Carpenter 2005b, 40):

Table 7.5: *Most common pattern of emergence of the four main social-cognitive skills with the percentage of participants who passed each skill*

Chimpanzees

Imitative learning	Attention following	Communicative gestures	Joint engagement
67	100	100	0

Typically developing infants

Joint engagement	Communicative gestures	Attention following	Imitative learning
100	100	96	96

Children with autism

Imitative learning	Joint engagement	Attention following	Communicative gestures
92	75	67	58

The conclusion based on these data is that (Tomasello and Carpenter 2005b, 39–40)

chimpanzees' pattern resembled that of children with autism more closely than that of typically developing infants—in fact, with the exception of the skill of joint engagement, which no chimpanzee passed, it was identical to that of children with autism. For infants, joint engagement and communicative gestures ... led the developmental sequence whereas for the chimpanzees communicative gestures (always imperatives) was at the end and joint engagement never occurred.

---

115 On the impairments in children with autism see (Carpenter and Tomasello and Striano 2005), (Carpenter and Pennigton and Rogers 2001; 2002) and (Liebal *et al.* 2008).

Tomasello's mutual comparison of typically developing human infants, children with autism, and of great apes shows, if one takes into account also the data in Table 7.5, that typically developing human infants display the richest social-cognitive skills. The result of this comparison can be represented as follows:[116]

Figure 7.5: Comparison of social-cognitive skills

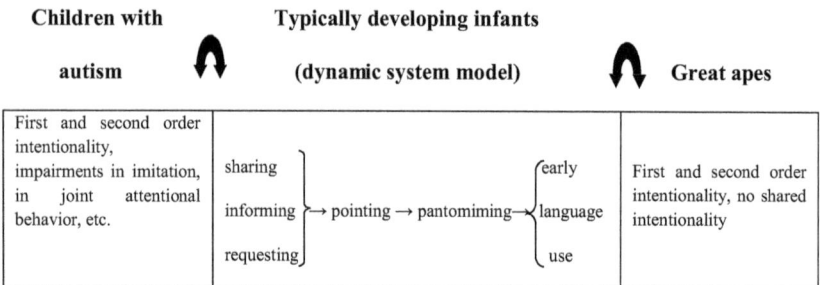

Children with autism	Typically developing infants (dynamic system model)	Great apes
First and second order intentionality, impairments in imitation, in joint attentional behavior, etc.	sharing ⎫ informing ⎬ → pointing → pantomiming → ⎧ early language requesting ⎭ ⎩ use	First and second order intentionality, no shared intentionality

Thus, because Tomasello deals here with the developmental constitution (genesis) of the whole structure of the social-cognitive skills, one can speak here about the *structural-genetical method* of research/explanation. By what type of entities is this method fulfilled? If one takes, for example, a typically developing infant and views it as displaying the most intricate structure, then in such a way understood infant can be viewed as a kind of a thought-object, namely, as a *classical thought-type*. If one takes also into account that Tomasello brings in for the purpose of comparison also autistic children and great apes, one can regard a typically developing infant, an autistic children, and a great ape as an *original thought-type in its own right*.

C) *The Structural-Historical Method and the Historical-Genetical Method*

In the works of Tomasello, especially in Chapter 5 of (Tomasello 2008) as well as in (Tomasello 2011), (Tomasello 2012), (Tomasello et al. 2012) and (Tomasello 2014), one can discern another method of research/explanation which I label the "structural-historical method" and which I shall try now to reconstruct. By means of this method one should be able to give an "explanation of evolution of cooperative communication between humans" (Tomasello 2008, 170), and where under "evolution" one should understand *phylogeny*. This method is,

---

116   The term "dynamic system model" is used in (Tomasello 2008, 135). The arrows indicate the directions of comparison.

on the one hand, related to and in fact developed from, the structural-genetical method, and should from the point of view of this relation provide "a kind of general primate starting point that might serve to isolate the evolutionary unique features of human social cognition" (Tomasello et al. 2005, 684). On the other hand, as I shall show, this method is, or should be, to be more precise, related to yet another method that I label as *historical-genetical* method, but which, to my knowledge, is absent in Tomasello's works. As I will show, this absence has its roots in the very nature of the subject-matter of his research.

Tomasello's approach to the phylogeny of human communicative cooperation is based on a thought movement which I label "retrospective" and by means of which he assigns to the stages of ontogeny of human infants'/children's cooperative communication, already reconstructed by means of the structural-genetical method, stages in the phylogeny of the human species; the former serves as a kind of an "conceptual key" for the latter. At the same time, he grounds this retrospective by bringing in data about great apes and children with autism. So, for example, he declares at the very beginning of his treatment of phylogeny of cooperative communication (Tomasello 2008, 170–171):

> The intimate relation between collaborative activities and cooperative communication is most readily apparent in the fact that they both rely on one and the same underlying infrastructure of recursively structured joint goals and attention, motivations and even norms for helping and sharing, and other manifestations of shared intentionality. This common infrastructure is most clearly evident in the fact that great apes have noncooperative forms of both group activities and intentional communication, underlain by skills for understanding individual intentionality, whereas human infants develop cooperative forms of both collaboration and communication underlain by skills and motivations for shared intentionality … We thus believe that … the common infrastructure of shared intentionality underlying both collaborative and communicative activities of contemporary human beings provides us with a tangible stamp of their common evolutionary origin.

And after dealing with that phylogeny, he states that "We have structured our evolutionary account of human cooperative communication around the emergence of the three major motives involved: requesting, informing, and sharing" (Tomasello 2008, 243), that is, his *conceptual reconstruction of the phylogeny of cooperative communication draws explicitly on the conceptual grasping of its ontogeny*.

This dependence of the conceptual reconstruction of phylogeny on the conceptual reconstruction of ontogeny, as well as the conceptual retrospective shift from the latter to the former, is viewed by Tomasello as in fact necessary because phylogenetic "stories are always woefully undetermined by data" (Tomasello and

Carpenter 2005c, 119). He labels such stories as *quasi-evolutionary tales* (Tomasello 2008, 198, 235), since they are more than just-so-stories because they can reconstruct "certain logical or at least plausible ordering relationships among the various components of human communication" (Tomasello 2008, 191). He acknowledges, however, that "We do not have a specified and detailed evolutionary story to tell" (Tomasello 2008, 191), and "We have not focused at all here on asking when particular things happened during human evolution; we have chosen, rather, to focus simply on the ordering of events" (Tomasello 2008, 235).

Tomasello starts his retrospective thought-movement by reflecting on group activities of chimpanzees, such as hunting, which he interprets as acts without prior synchronization, joint goal, and in which the chimps do not play different and complementary roles; even in cases when parallel roles are played (e.g., in simultaneous pulling), no communication was observed among them (Tomasello 2008, 175–176). At the basis of all these facts, he concludes, should be that great apes as individual intentional agents "have neither skills nor the motivations to form with other joint goals and joint attention, or otherwise participate with others in shared intentionality" (Tomasello 2008, 177).

In a subsequent step, Tomasello moves conceptually to the reconstruction of the main features of human-like collaborative activities characterized by an intentional structure involving joint goals and complementary roles; they are viewed as the *sine qua non* of human collaborative actions (Tomasello 2008, 181). In order to move conceptually from the description of great apes' intentional action to that of human cooperative communication, he proposes a sequence based on the description of the following three basic motives of that communication (Tomasello 2008, 192):

- to explain the granting of requests and the initial motive to help by informing, we invoke mutualism (the request is granted or the information is offered because it helps us both);
- to explain offering help by informing outside of mutualistic contexts, we invoke reciprocity and indirect reciprocity (help is offered because it adds to my reputation for cooperativeness so that others will want me as a cooperative partner—and help me in return); and
- to explain sharing emotions and attitudes, we invoke cultural group selection (emotions and attitudes are shared as a way of increasing common ground and solidifying group membership).

For the first motive, it holds that it could have begun initially with individuals becoming more tolerant, generous, and less competitive with one another (especially in feeding contexts). Here, Tomasello uses as the conceptual point of departure the behavior of chimpanzees who display such features in interaction with humans. So, in this step the claim is that "in human evolution greater tolerance among conspecifics would have been enough to begin moving in the direction of true collaboration as well as imperative pointing—with no further cognitive skills necessary beyond those of modern-day great apes" (Tomasello 2008, 194). At the same time he views *pointing* as the most obvious candidate for requesting/offering help. As another step, he suggests that (Tomasello 2008, 194)

> these individuals, who are coordinating actions with one another more regularly and tolerantly, would then be in a position for natural selection ... to specifically favor cognitive and motivational machinery supporting more complex collaborative interactions.

For the second motive, Tomasello brings in the fact that contemporary humans also request/offer help outside of mutualistic collaboration. So, in order to give a generalization of his reflections about the first motive, he brings in the concept of indirect reciprocity understood as a situation "in which individuals choose to help to cooperate with others who have good reputation for helping and cooperating in general" (Tomasello 2008, 200).[117] As an adequate mean of communication here he views in addition to pointing also *iconic gestures*.

In the case of the second motive, as well in the case of iconic gestures, Tomasello draws from reflections on great apes' choice of more skillful than less skillful cooperators in instrumental actions,[118] while "iconic gestures ... presumably derive from ape intention-movements—incipient real actions—but add a representational dimension based on simulation/imitation and the recipient's comprehension of the communicative intention" (Tomasello 2008, 204).

Finally, in the third motive, Tomasello locates behavior that conforms to the behavior of the conspecifics; this conformity with the behaviors and actions of others is usually achieved by imitation, while the preferred form of communication here could be *expressive-declaratives*.

In a later step, Tomasello deals with the emergence of conventional communication that in his view developed from iconic gestures and corresponds to infant's cognitive development from iconic gestures to linguistic symbols. So, the phylogeny of human cooperative communication could have followed the following scenario (Tomasello 2008, 235):[119]

---

117 On indirect reciprocity in apes see (Melis – Hare – Tomasello 2006a).
118 On this see (Melis – Hare – Tomasello 2006b).
119 Here "demonstrative" stands for the most basic communicative act embodying a spatial component of distance from the speaker and is often accompanied by a pointing gesture, "deictics" stand for gestures used to direct the attention of a counterpart to

Figure. 7.6: Tomasello's quasi-evolutionary scenario of the genesis of human cooperative communication

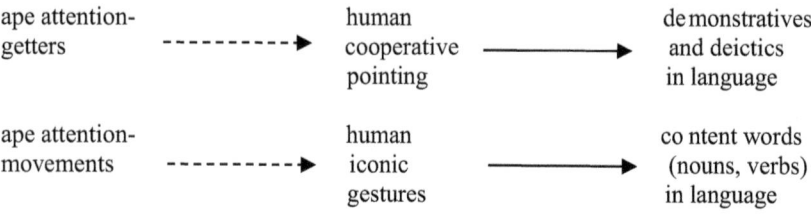

Based on these "quasi-evolutionary stories," Tomasello proposes the following table as representing the evolution of human cooperative communication (Tomasello 2008, 239):

Table 7.6: Tomasello's reconstruction of the phylogeny of human cooperative communication

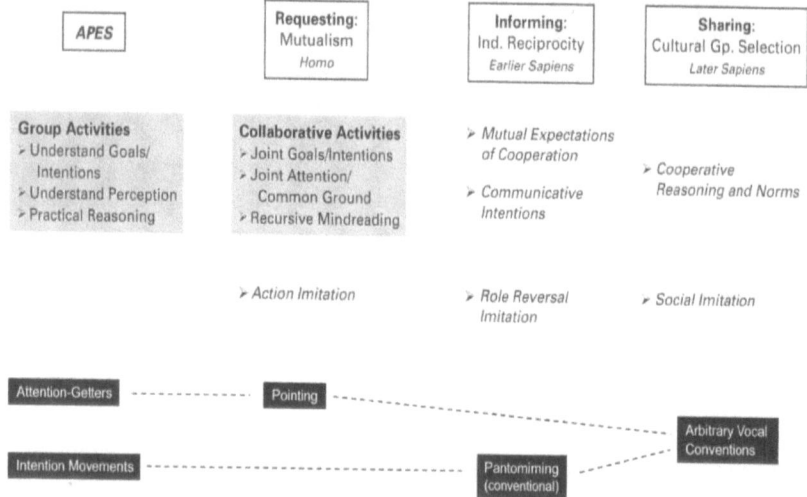

Tomasello states, with respect to the first column, that "modern-day great apes—representing our model of the starting point—have many of the necessary components for human cooperative communication" (Tomasello 2008, 238); they

---

outside objects, a content word is an element of language corresponding to an iconic gesture, while nouns and verbs are the most fundamental kinds of content word in a language.

understand goals/intentions as well as perceptions, and they are engaged in practical reasoning. Thus, at least in my view, he presupposes that the communicative skills in nowadays living great apes and in their extinct ancestors can be viewed as at least very similar or even as identical. Then, the term "Apes" in this table should stand for nowadays living great apes *and* their extinct ancestors.

Skills and motivations for shared intentionality appeared initially within mutualistic collaborative activities and were symptomatic of creatures labeled *Homo* (second column). Then, helping was generalized to other contexts based on indirect reciprocity; this happened in creatures labeled *Early Sapiens* (third column). Sharing attitudes as the third motive originated in the motive to be like other groupmates; it was symptomatic of *Late Sapiens* (fourth column). The bottom row in the table expresses the contention that apes' attention-getters evolved into human pointing, the latter was then transformed into iconic gestures, then came communicative conventions and, finally, a shift to the vocal modality took place (Tomasello 2008, 239–241).

Let me now deal with Table 7.6 from the point of view of methodology. On the one hand, it is, as shown already, *based on a unification of the conceptual reconstruction of the ontogeny of human infants/children with a conceptual characterization of nowadays living great apes* and, as I claimed, *on the supposition that the communicative skills in nowadays living great apes and in their extinct ancestors can be viewed as very similar or even identical*. This table thus draws explicitly on the results of application of the structural-genetic method. On the other hand, however, that table should delineate the hypothetical stages in *phylogeny*. This double characterization of Table 7.6 comes to the surface in Tomasello's reminder: "we see the terms *Homo, Early Sapiens, Late Sapiens* as handy and suggestive labels concerning evolutionary sequences, nothing more" (Tomasello 2008, 293). Thus, this table cannot be viewed as a truly evolutionary diagram for the genesis of human cooperative communication. In order to be such, the concepts *Homo, Early Sapiens,* and *Late Sapiens* from this table would have to be related to an evolutionary tree of *hominin species*.[120]

Table 7.6 should thus be viewed as an *in-between stage* which mediates between, on the one hand, the conceptual reconstruction of the ontogeny of typically developing human infants/children, of the cognitive/communicative skills of children with autism and those of great apes and, on the other hand, a highly

---

120 Here I understand under "a hominin" an individual belonging to a tribe that is composed of current humans and their direct and lateral ancestors that are not also ancestral to chimpanzees. If also the term "Apes" from Table 7.6 should find its counterparts in an evolutionary tree, then this tree should be that of *hominoids*.

developed evolutionary tree. The latter should deal with the historical genesis of species; the method applied in such a tree I label *historical-genetical method* and it should be viewed as the end-point of the thought movement that started from the structural-genetical method. So as from the point of view of methodology, Table 7.6 should stand for the *mediating link* between, on the one hand, *structural*-genetical method and the historical-*genetical* method, on the other hand, I label the method given in such a tree as *the structural-historical method*. The concepts obtained by the structural-historical method serve as means and guide for the investigation and research into the thought entities appearing in the historical-genetical method. What unifies both methods is the fact that they both stand for an attempt at a *historical explanation*. The latter I characterize, from the point of view of methods employed in theory construction, as *the process and result of a conceptual reconstruction of the formation of entities under investigation, where this formation belongs to the past, and where this explanation is accomplished in the framework and boundaries set by the research and explanatory problems to be solved.*[121]

In Tomasello's more recent works (Tomasello 2011), (Tomasello 2012), (Tomasello et al. 2012) and (Tomasello 2014) one can detect a shift of emphasis, at least in my view, in the order of conceptual reasoning. Now, the center stage is occupied by the above characterized three-phase evolutionary hypothesis, to which is assigned additional information drawn from various sources.

In Phase 1, in which the social cognitive capacities of a member of a species are hypothesized, Tomasello chooses as "[o]ur best, living models for this creature ... humans' closest primate relatives, the non-human great apes" (2014, 15). Its behavior, namely, that of an individual forager who never voluntary share the spoil of even such a hunt in which it collaborates with its conspecifics, is viewed by him as that of the last common ancestor of non-human great apes and of the various *Homo* species.[122] Its thinking should have displayed the following three key cognitive characteristics: i) the ability to operate in the mind with abstract cognitive representations like categories, schemas and models; ii) based on the latter to infer representation of causal and/or intentional situation which as yet do not exist; iii) self-monitoring of the decision making process.

In Phase 2, Tomasello draws on two sources of information. One is the prosocial behavior of pre-linguistic human infants based on pointing and pantomiming, and the second are the foraging activities of small-scale modern hunting

---

121 Here I draw partially on (Jaeck 1988, 228).
122 On this cooperation see (Boesch 2002).

groups.[123] These behaviors should correspond to the behavior of a *Homo* ancestor of our species; possibly that of *Homo heidelbergensis* (2014, 40). The latter could have displayed a joint- or *we*-intentionality characterized by cognitive representations (perspectival, symbolic); inferences (socially recursive) and self-monitoring in the sense of regulating his/her own action from the point of view of the perspective of a cooperative partner (2014, 33).

Finally, the passage to Phase 3 was, according to Tomasello, trigged by the population growth in, and competition between, these groups.

From the point of view of the methods employed, it is worth noting that he views his three-phase hypothesis as "a very general evolutionary story (supported by comparative experimental data)" (Tomasello et al. 2012, 675) and also as a "an evolutionary fairy tale" (2011, 39). The introduction of Phase 2 into that hypothesis as a *mediating link* between Phase 1 and Phase 3 he justifies as follows: (2014, 33):

> Assuming ... that great apes are representative of humans' last common ancestor with other primates, then it would seem that we need an intermediate step in our natural history. We need some early humans who were not yet living in cultures and using languages, but who were nevertheless much more cooperatively inclined than their last common ancestor.

And he also states the following: "It was necessary because one cannot even imagine going directly from ape-like competitive interactions and imperative communications to modern human culture and language with no evolutionary immediacy" (2014, 150).

---

123 On this see (Hill and Hurtado 1996) and (Hill 2002).

Chapter 8: Tool use by Chimpanzees (*Pan Troglodytes*): New Conceptualization and a New Measure for Quantification

8.1 Introduction

The claim, attributed to Benjamin Franklin, that tool-making is a uniquely human activity has gradually lost its persuasiveness in the last fifty years or so. Recent research indicates that chimpanzees (*Pan troglodytes*) produce and employ complex tools, and even utilize them for hunting (Pruetz and Bertolani 2007; Wilfried and Yamagiwa 2014).

My aim, based on these facts, is to provide a new conceptualization of tool use by chimpanzees. The starting point will be data on chimpanzee tool use and food sharing. I shall then compare these data with the now widely accepted conceptualization of that use in (Matsuzawa 1996) and well as with the conceptualization given in (Hayashi and Mizuno and Matsuzawa 2005) in order to propose an alternative explanation, by drawing on the conceptualization of tool use and consumption of its product by humans in order to develop a set of concepts suitable for the description of chimpanzee tool use. Finally, I shall propose a new measure for chimpanzee tool use based on this conceptualization.

8.2 Tool Use in Chimpanzees

In my analysis, I shall focus on three kinds of tool use by chimpanzees: leaf sponging, honey gathering, and termite fishing.

8.2.1 Leaf sponging

According to (Sanz and Morgan 2010), this activity is based on the use of a leaf sponge used to extract water from a tree basin. Here a chimpanzee acts by means of a hand on a plant, the product being a leaf torn away. Then, the chimpanzee by means of a hand stuffs the leaf into the mouth, where it is chewed and then taken out by a hand – the product being a modified leaf. Next, the chimpanzee puts the leaf by means of a hand into a tree basin, the product is now a water soaked leaf. Finally, the chimpanzee puts the leaf by mean of the hand into the mouth and extracts the water.

203

8.2.2 Honey extracting

In (Boesch and Head and Robbins 2009; Fowler and Sommer 2007; McLennan 2011; Pascual-Garrido et al. 2012; Sanz and Morgan 2007; Sanz and Morgan 2009; Sanz and Morgan 2010; Wilfried and Yamagiwa 2014) honey extraction is described as follows. In the case of an arboreal beehive, the chimpanzee by means of a hand rips off a branch from a small three; the product is a stout stick, which is then employed manually as a pounder to break open the protection of the hive's entrance. Once the product, the beehive broken open, is obtained, the chimpanzee employs another means, namely, a collector (fluid dip) – having previously been torn from a plant – to dip (scoop) the honey out of the hive. Finally, the chimpanzee shifts the collector to its mouth and consumes the honey.

In the case of subterranean beehives, a stick together with a hand is used as a perforator to open the ground and then a piece of bark is used as a swabber to dip the honey out.

8.2.3 Termite fishing

In (Deblauwe et al. 2006; Sanz and Morgan and Gulick 2004; Sanz and Call and Morgan 2009; Sanz and Morgan 2010; Sanz and Morgan 2011) two different tool sets for termite fishing are described. For subterranean nests a hand-held stout stick is first produced from a branch torn from a tree and used to puncture the nest. The termites are then fished out by means of a probe. For epigeal mounds, a thinner twig is first torn away from a plant and then used to reopen holes used by swarming termites. In this case, too, a fishing probe is used to extract the termites.

8.3 Chimpanzee Food Sharing

While research into chimpanzee tool use yields a gradually increasing stock of information confirming this behavior among most of the chimpanzee subspecies in West, Central, and East Africa (for the latter see also Sherrow 2005), the data on food sharing in chimpanzees yield a rather diffuse picture.

While meat sharing is widely distributed in male-male interactions (Mitani and Watts 2000; Mitani and Meriwether and Zhang 2000) primarily for the purpose of coalition formation, transfer of non-meat food and tools is rare (Pruetz and Lindshield 2012). When it is observed in male-female interactions, the primary reason seems to be mating strategies (Hockings et al. 2007), even if not for the short term (Gilby et al. 2010), but still for the long term (Gomes and Boesch 2009). As another reason for food sharing seems to be as a strategy for escaping harassment (Crick et al. 2013; Gilby 2006; Stevens 2004).

All those data were obtained by observing wild chimpanzees; experiments on captive chimpanzees showed, on the one hand, a clear tendency not to share food with conspecifics and to feed alone (Bullinger et al. 2013), but, on the other hand, pro-social sharing behavior among nonkin (Horner et al. 2011).

8.4 The Matsuzawa Model

T. Matsuzawa presented in (1996; 2001) a conceptualization of chimpanzee tool use based on the identification of detached objects involved in this tool use. These objects are viewed as being unified in a cluster, the latter represented graphically according to the temporal order in which they are interconnected in tool use. Thus, for example, once a nut is placed on a stone anvil, and then the nut on this anvil is being hit by a hammer stone, the graphical representation is as follows:

Figure 8.1: T. Matsuzawa's representation of a two-level tool use involving three objects

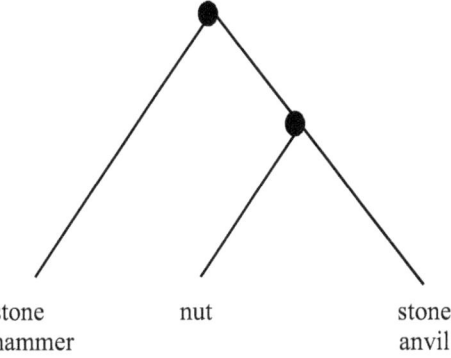

Here the points given by the intersection of the lines stand for a nod, while the higher level nod stands for the interaction of the nut-anvil cluster with the hammer stone.

Despite the innovative features of this conceptualization, it displays the following three deficiencies. First, it leaves out the actor who uses the tools; as I will show below, by conceptually taking into account the presence of the actor, a fundamental difference between chimpanzees' and humans' tool use can be conceptualized. Second, it leaves unmentioned the participation of parts of the agent's body in the tool use. Even if they – using the terminology of Parker and Gibson (1977) – belong to agent's anatomical equipment, still they can serve as tools when, for example, separating an object from a substrate. A hand or teeth are used by a chimpanzee to separate a twig from a plant which, then, in turn, is used for termite fishing or honey extraction. Thirdly, it leaves out the consumption of the final product resulting from the tool use. What thus drops out from the conceptualization is the cycle composed of three components: the agent, the tool use, and consumption of the product resulting from the tool use. The conceptualization of this cycle makes it possible, as I will show below, to understand the differences between tool use by chimpanzees and tool use by humans.[124]

8.5 The Hayashi – Mizuno – Matsuzawa Model

The Matsuzawa model was further developed in (Hayashi and Mizuno and Matsuzawa 2005; Hayashi and Takeshita and Matsuzawa 2006). So, for example, when a chimpanzee by employing its right hand puts a nut on a stone A with its right hand and then hits the nut with stone B by using its right hand, the first action is conceptually approached via the expression (leaving out the name of the agent) "*put a nut on a stone with one hand*," while the second one via the expression "*hit the nut on a stone with one hand*." In the first expression the underlined sub-expressions are interpreted in such a way that "*put*" should stand for the *action*, "*nut*" for the *manipulated object*, "*stone*" for the *location* in the sense of the *target of the manipulated object*, and "*one hand*" for the *body part that is performing the manipulation*. In the second expression "*hit*" stands for the *action*, "*nut on a stone*" for the *location*, "*stone*" for the *manipulated object*, and "*one hand*" for the *body part that is performing the manipulation*.

Even if this approach already takes into account, in contradistinction to the Matsuzawa model, both the action and the hand performing the tool use, still the following two objections can be stated against it.

First, like the Matsuzawa model, it leaves out the both the agent and the consumption of the final product of tool use by this actor.

Second, the concepts *location* and *object of manipulation* do not take into account that both refer in these examples to manipulated entities which, however, at the same time fulfill different roles in this tool use. In the first case, the nut is the entity which is manipulated in this action – I will refer to it by the term "object of tool use" or "object" for short– by employing a hand which is also a manipulated entity. The latter I label as "means of tool use" or "means" for short, while the placement of the nut on the stone is the state of affairs produced by this

---

[124] By including the concept of consumption I follow Marx's views presented in the introduction to the manuscript *Grundrisse*.

action; hereafter, I will refer to it by the term "product of tool use" or "product" for short.[125]

In the second case, both stones *A* and *B* can be viewed as means which I, together with the hand(s) employed, label in the unity in which they appear here, following (Boesch 2013), as a "combined means." Finally, the nut cracked open I label as "product of tool use," where "tool use" refers here to the pounding of the nut placed on stone *A* by means of stone *B*.

The concepts labeled as "agent," "means," "object," "product" and "consumption of the final product" enable to provide a new conceptualization of tool use as performed by chimpanzees.

8.6 A New Conceptualization

The ambition of contemporary research into chimpanzee tool use is to provide a set of concepts grasping the main characteristics of chimpanzee tool use. This set should then serve the aim of comparing human tool use with chimpanzee tool use and thus "resolve questions of what is specific about human tool use" (Boesch and Head and Robbins 2009, 568). This aim, from the point of view of conceptualization, can, of course, be achieved only if we already have at our disposal a set of concepts which grasp the main characteristics of human tool use. By comparing these two sets of concepts, the initial set of concepts stated for human tool use could then be made more dense and precise. The idea behind all this is that we can somehow provide a set of concepts for chimpanzee tool use independently from the set of concepts describing human tool use.

The entirety of research into chimpanzee tool use is driven, first, by the acknowledgment that the human species is *the* tool making species and, second, by a certain, pre-understanding what this tool use stands for. This pre-understanding is, however, never dealt with explicitly in articles on chimpanzee tool use. It thus remains rather tentative, not explicit, and conceptually not precise enough.

Moreover, in contemporary research it is widely acknowledged that human tool use displays a more complex structure than that of chimpanzees. So, priority should be given to an explicit and precise conceptualization of tool use by humans, from which a set of concepts suitable for describing tool use by chimpanzees should then be derived. Stated otherwise, *the set of concepts used for the description of tool use by humans should serve as the key and starting point for*

---

125 Here I employ the terminology developed by Marx in Chapter "Work Process und Valuation Process" from *Capital*, Volume 1.

deriving, by means of a reduction driven by data on chimpanzee tool use, a set of concepts which would help us to understand chimpanzee tool use.

Earlier, I introduced the terms "agent," "means," "object," "product," and "consumption of the final product." In order to employ them for describing features specific to human tool use, I unify them with the term "shared" used in (Tomasello et al. 2005) and the term "collective" used in (Tomasello et al. 2012; Tomasello 2014), both of which are useful for the characterization of intentionality which is unique for humans, namely, "shared (collective) intentionality."

What is unique to tool use by humans is not that they employ means on objects to obtain products which they, finally, consume, but that the tool use in which they are engaged is by its very nature collectively shared, in the following threefold sense. *First, a product of tool use by a human being becomes the means and/or object of the tool use by another human being. When the product of a human being's tool use is employed by another human being as means for action on objects, then that product acquires in the action of the latter human being the status of a tool; thus, the tool is produced via cooperation.*

Second, the product of human tool use, into the production of which entered the products of tool use by other human beings, is shared in the sense that it is collectively consumed. Human tool use is collectively shared because it is reproduced as such only by means of a shared consumption of its final product, without which the tool use itself could not be reproduced.

Thus, third, the human agents who are involved are reproduced in the network of relations of tool use and the consumption of the results of tool use as cooperated beings.

This threefold characterization of human tool use I represent for three human beings as H_1, H_2, and H_3 in Figure 8.2.

Figure 8.2: Cycle of cooperated tool use and shared consumption of its product

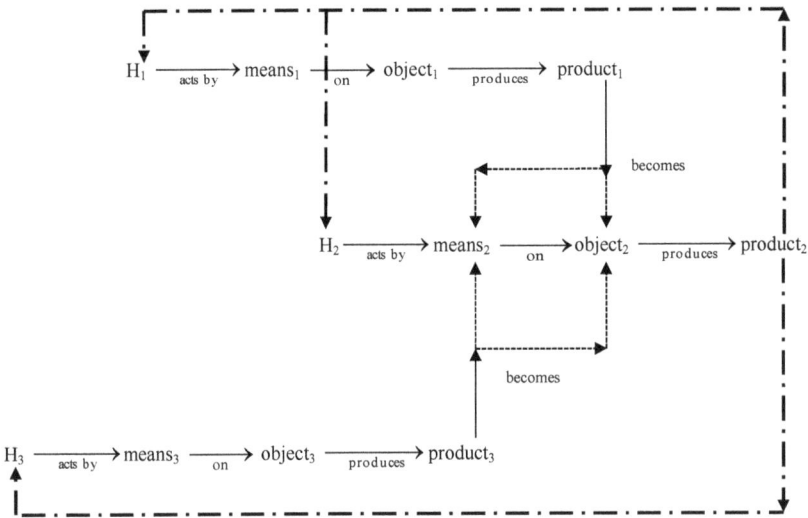

In order not to overburden the figure, I do not express in it the possibility that the products of H_1's and H_3's tool use need not enter entirely into the tool use of H_2, but can be partially consumed by H_1 and H_3.

The thin dotted lines indicate the paths along which products of tool use, in the sense of an employment of means on objects, turn into means and/or objects in another tool use. The bold dotted lines indicate the shared consumption of the final product by H_1, H_2, and H_3. The shared consumption of the end-product of a collective tool use is at the same time the starting point for the renewal of this use.

This conceptualization of human tool use provides a characterization of its core features, from which, of course, particular cases may sometimes deviate. Thus, one can imagine its reduced form when, for example, a particular human being proceeds, in order to quench his/her thirst, in a manner similar to that of a chimpanzee, namely, by leaf sponging.

Thus, by applying the abovementioned threefold characterization of human tool use and consumption of its products to the available data on tool use and consumption of its products by chimpanzees, one can understand the core characteristics of chimpanzee tool use and consumption of its products. Thus, even if chimpanzees share with nonkin food, both meat and nonmeat by nature, as well as tools, they never engage in a collectively shared tool use in the sense that the products of a chimpanzee's tool use enter as means and/or as objects into the tool use by another chimpanzee.

Furthermore, they are not engaged in shared consumption of products of a collectively shared tool use. What they are engaged in, instead, is individual tool use, whose products are shared either frequently, as it the case with meat, or less frequently, as it is the case with non-meat food. What sometimes is subject to sharing by chimpanzees are also tools; here, however, the products of their employment are – as far as my knowledge of the data goes – not collectively shared.

Based on the conceptualization of the core characteristics of human and chimpanzee tool uses, while taking into account possible individual deviations from core characteristics of human tool use, their relation can be expressed by means of Table 8.1.

Table 8.1: *Types of tool use and consumption of products of tool use by chimpanzees and humans*

Type of consumption \ Type of tool use	individual	shared
individual	chimpanzees & humans	chimpanzees & humans
shared	chimpanzees & humans	humans

The fact that the product of a chimpanzee's tool use does not acquire the status of a means of action on objects in the activity of another chimpanzee means that chimpanzees *do not produce tools via cooperation*. The lack of such a type of cooperation can serve as the basis for the explanation of a feature which is, when compared to humans, specific for chimpanzee tools, namely, their rather low degree of complexity.

The most complex tool known to be used by chimpanzees is a three- or four-stone combined tool: a stone hammer combined with a stone anvil and one or two stone wedges supporting the anvil (Biro and Carvalho and Matsuzawa 2010; Carvalho et al. 2008). Completely lacking among tools produced and used by chimpanzees are the so-called *composite tools* (Boesch 2013), that is, tools produced of at least two different materials that are joined together so as to function as one tool.

Given the lack of cooperation in tool production, the latter is – from the point of view of the individual chimpanzee – subjected to a certain time limitation. The more complex the tool – given that lack – the more time does the individual tool manufacturer need to spend in its production; and beyond a certain time period that production loses its manageability for this individual with respect to the product it should yield. Contrary to this, once tool production is cooperative, then the time-span in which a tool-using individual should participate, in order to obtain the respective result to be consecutively shared, can remain manageable for this very individual.

The lack of cooperation in tool production by chimpanzees also can serve as the key for the understanding of the fact, expressed in (Kitahara-Frisch 1993) by means of the terminological differentiation between a "primary tool" and a "secondary tool," that chimpanzees do not, in order to use a (primary) tool, employ another (secondary) tool for its production. In an instance of cooperated tool production the product of an agent's tool use turns into a secondary tool in the activity of another agent whose activity yields a product which, in turn, in the activity of yet another agent is employed as a tool in the sense of a means of action on an object.

8.7 A New Measure for Chimpanzee Tool Use

Let me now introduce a measure which would make possible a quantification of chimpanzee tool use and consumption of its final product.

This measure is based on a matrix assigned to that use and consumption, so that in each column what is given in the first line is the number of actors involved in tool use, A; in the second line the number entities making up the tool, T; in the third line the number of entities making up the object, O; and in last line the number of actors, C, consuming the final product. Each column stands for one sequence of an agent's acting by means of a tool on an object and final consumption (if performed); schematically:

$$\begin{bmatrix} A_1 & A_2 & A_3 & \dots \\ T_1 & T_2 & T_3 & \dots \\ O_1 & O_2 & O_3 & \dots \\ C_1 & C_2 & C_3 & \dots \end{bmatrix}$$

Once this matrix is available, then the measure for the tool use is given by the sum total over columns and lines of all numbers in this matrix.

As an example, let me compute the measure for chimpanzee leaf sponging. By combining Figure 8.2 with the conceptualization of the core features of chimpanzee tool use and consumption, the diagrammatic representation for a chimpanzee's leaf sponging is as follows ("Ch" stands for a chimpanzee):

Figure 8.3: Cycle for leaf sponging and consumption of water by a chimpanzee

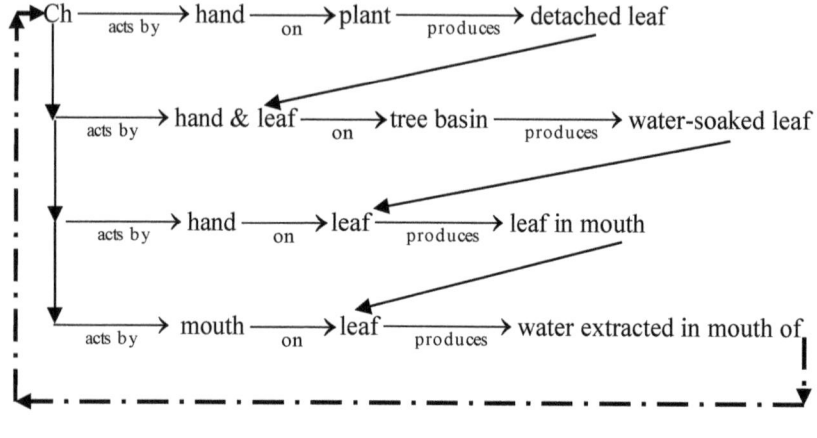

The matrix corresponding to this figure is $\begin{bmatrix} 1000 \\ 1211 \\ 1111 \\ 0001 \end{bmatrix}$. The zeros in the first line express the fact that just one chimpanzee is involved in all four sequences of tool use. The zeros in the last line express the fact that only the final product is consumed and not the intermediate product (e.g., when two leaves would be detached instead of one, but only one would enter into the next sequence, while the remaining one would be consumed by the chimpanzee). The numeral "1" in the last line expresses that just one chimpanzee is involved in the consumption of water.

The measure for tool use as represented by that matrix is 11. It can thus be readily seen that the more complex a tool use is – due to the increased number of actors, entities involved as a tool and as an object, as well as due the increased number of actors consuming the final product – the higher the number assigned to it as its measure.

By way of summary and conclusion, I shall enumerate some of the benefits and uses of such a characterization. It can be used to measure inner-group as well inter-group differences in one subspecies, as well as between subspecies of chimpanzees. It can also be used to measure the dynamics of tool use in the ontogenesis of chimpanzee infants. In addition, it can be employed to measure the dynamics of tool use of a whole group due to the impact of various external influences (e.g., interaction with humans). Finally, it can be used to compare the complexity of *wild* chimpanzee tool use and consumption of its final product with that of their *captive* counterparts.

Part III:
Methods of Theory Construction in Political Economy

Chapter 9: Marx's Method of Theory Construction: Categories, Magnitudes and Laws

The aim of this chapter is to reconstruct the conceptual methods used by Marx in the construction of Volume I of *Capital* as well in the manuscripts of its Books II and III.

The circumstances enabling such a reconstruction are now more favorable than ever before, especially since the conclusion of the publication of section II of the complete edition of Marx and Engels – known under the abbreviation *MEGA*. We now thus have at our disposal not only the reprints of all the editions of the first volume of *Capital* authored or authorized (1983; 1987; 1989) by Marx, but also of the manuscripts he intended for *Capital*'s Book II (1988b; 2008; 2012a; 2012b) and Book III (1992; 2003a; 2012c).

Drawing on these textual resources I can now identify the key components of the conceptual methods he employed in his economic works, allowing thus to overcome wrong, distorted or incomplete views of these methods.

In modern economics, Joan Robinson is a suitable representative of such a view. I shall start, therefore, with an analysis of her approach and evaluation of Marx's economics and show how they are rooted in an incorrect understanding of the structure of *Capital* and thus, in fact, of the methods he employed in the construction of the *Capital*.

And, in contemporary philosophy of science, Leszek Nowak, who in (1980) made an attempt at a reconstruction of the methods of idealization and concretization employed by Marx in *Capital*, is an apt representative of this position. Despite its indisputable positive aspects, his analysis fails in some crucial regards in the reconstruction of the method employed by Marx.[126]

Accordingly, in order to overcome the deficits in Robinson's and Nowak's efforts, I shall first reconstruct the principal methods employed by Marx in *Capital*, both in Volume I and in the manuscripts of Books II and III. Then, I shall present a general characterization of these methods.

In order to prevent any possible misunderstanding, I would like to emphasize that in this chapter I do not deal with all categories of political economy treated by Marx in Volume I of *Capital* and his manuscripts of Book II and Book III. I

---

126 On these positive aspects see (Hanzel 1999).

leave out, for example, the categories of absolute surplus value, relative surplus value, agricultural rent, merchant profit, and enterprise profit. Instead, I shall focus on methods employed by Marx in his introduction and treatment of categories of political economy.

All quotes from Marx are my translations of texts published in the *MEGA*; quotes from Hegel are my translations from the German.

9.1 Joan Robinson on Marx's Concepts of Value and on the Relation of Volume I to Volume III of Capital

In one of her later works, Robinson gives quite a positive evaluation of Marx's economic oeuvre. She states the following (1979, ix–x):

> The forces which govern the distribution of the product of industry between wages and profits are the central features of a capitalist economy. Marx was quite right, quite apart from ideology, to put the rate of exploitation into the center of the picture … Thus it seems to me that, whether it is expressed in terms of *value* or not, the Marxian theory provides the best foundation that is available for a system of analysis to be elaborated and applied to actual problems.

But even in this positive evaluation one can detect her reservations about Marx's employment of the category of *value*. From the point of view of this chapter it is worth dealing with these reservations. As early as in her 1942 essay on Marx she states that (1966, 22):

> no point of substance in Marx's argument depends upon the labour theory of value. Voltaire remarked that it is possible to kill a flock of sheep by witchcraft if you give them plenty of arsenic at the same time … Marx penetrating insight and bitter hatred of oppression … supply the arsenic, while the labour theory of value provides the incantations.

In the second introduction of 1966 to this essay she evaluates Marx's employment of the category of value as follows: "The concept of value seems to me a remarkable example of how a metaphysical notion can inspire original thought, though itself it is quite devoid of operational meaning" (1966, xi).

This claim about the metaphysical status of the category of value obviously draws on the views of Popper on metaphysics (1964, 8–9):[127]

> The hallmark of a metaphysical proposition is that it is not capable of being tested … It purports to say something about real life, but we can learn nothing from it. Adopting Professor Popper's criterion for propositions that belong to the empirical sciences that

---

127 For a development of these views see (Watkins 1958) and (Watkins 1975).

they are capable of being falsified, it is not a scientific proposition. Yet metaphysical statements are not without content ... Metaphysical propositions ... provide a quarry from which hypothesis can be drawn. They do not belong to the realm of science and yet are necessary to it. Without them we could not know what is it what we want to know.

She states also that "[o]ne of the great metaphysical ideas in economics is expressed by the word 'value'" (1964, 29). This must mean, since Marx applies the category of value to labor power, that "[o]n this plane, the whole argument appears to be metaphysical; it provided a typical example of the way metaphysical ideas operate. Logically it is a mere rigmarole of words" (1964, 39). And in a similar manner she declares that "[t]he theory of *value*, in the narrow sense of a theory of relative prices, is not the heart of Marx's system ... and nothing that is important in it would be lost if *value* were expunged from it altogether" (1950, 360)

Also worth mentioning here are her views on the relation of Volume I to Volume III of *Capital*, having here in mind of course that with respect to the latter volume she had at her disposal only Engels' edition of Marx's manuscripts intended by the latter for Book III of *Capital*. She declares about the role of the category of value with respect to that relation the following: "It seems, certainly, perplexing, as we follow the uphill struggle of Marx's own mind from the simple dogmatism of the first volume of *Capital* to the intricate formulations of Volume III But if we start from the vantage ground of Volume III the journey is much less arduous" (1966, 10).

And she makes the following salient claim about that relation, worth quoting in its entirety (1966, 14–15):

> The main problem ... Marx does not attempt to deal with in Volume I at all. This concerns the tendency of the rate of profit to equality in different lines of production. In a system in which prices correspond to *values*, the net product of equal quantities of labour is sold for equal quantities of money. Thus (given uniform money wage rate) surplus, in terms of money, per unit of labour is everywhere equal. To say that relative prices correspond to relative values is the same thing as to say that the rate of exploitation is equal in all industries. But if capital per man employed (the organic composition of capital) is different in different industries, while profit per man (the rate of exploitation) is the same, profit per unit of capital must vary inversely with capital per man. It would be possible for both the rate of profit and the rate of exploitation to be equal in all industries only if the ratio of capital to labour employed were also equal. In Volume I, Marx leaves this question open. In Volume III, he shows that the capital per man varies with technical conditions, while competition between capitalists tends to establish a uniform rate of profit. The rate of exploitation therefore cannot be uniform, and relative prices do not correspond to *values*.

The final diagnosis she then gives with respect to the relation of Volume I to Volume III of *Capital* is as follows: "As I see it the conflict between Volume I and

Volume III is a conflict between mysticism and common sense. In Volume III common sense triumphs but must still pay lip-service to mysticism in its verbal formulations" (1966, 15).

As can be readily seen from the last quote, Robinson *interprets and evaluates Volume I retrospectively in light of Volume III*, that is to say, *backwardly from the point of view of the network of categories of political economy derived by Marx in Volume III*. The same retroactive procedure is explicitly applied on the methodological level by Leszek Nowak in his (1980) in which he made an attempt at a reconstruction of the structure of the law of value in Marx's *Capital*

9.2 Leszek Nowak on Marx's Method

The starting point of Nowak's reconstruction is the law of value which he understands as relating the prices of commodities with their value, so that the former are formed with respect to the latter (1980, 3). But this relation holds only under certain conditions. So, for example, Marx states "*if* supply and demand equilibrate each other, the market prices of commodities will correspond ... with their values, as determined by the respective quantities of labour required for their production" (2003b, 166). This condition Nowak views as an idealization and the law of value which he reconstructs as a conditional with the following structure:

If Cx and $Dx - Sx = 0$, then $Px = l(Vx)$

Here "Cx" states that x is a commodity, "$Dx - Sx = 0$" stands for the idealization that the difference between the demand for x and supply for x is equal to zero, "Px" stands for the price of x, "Vx" for the value of x and "l" denotes a linear function.

Nowak identifies in total the following eight idealizations under which the equation $l(V)$ should be valid (1980, 4–9), each of which I designate as $p_i = 0$ ($i = 1, 2, ..., 8$).

1. No obstacles exist for the free flow of capital and labor force, that is, commodity x is produced in conditions of perfect competition; $p_1 = 0$.
2. The difference between the exports and imports of the economy where x is produced equals zero; $p_2 = 0$.
3. The difference between the rate of surplus value characteristic of the capital employed in the production of x and the rate of surplus value in the economy where x is produced equals zero; $p_3 = 0$.
4. The difference between the average organic composition of agricultural capitals and the average organic composition of the remaining capital in the economy where x is produced equals zero; $p_4 = 0$.
5. The merchant profit on x is equal to zero; $p_5 = 0$.

6. The difference between the average organic composition of capital in the branch where x is produced and the average organic composition of the capital in the whole economy where x is produced equals zero; $p_6 = 0$.
7. The difference between the supply of x and the demand for it is equal to zero; $p_7 = 0$.
8. The difference between the organic composition of capital employed in the production of x and the average organic composition of capital in the branch where x is produced equals zero; $p_8 = 0$.

The law of value, given these eight idealizations, can be viewed as conditional with the following structure ("$p_{1-8} = 0$" stands for the conjunction of idealizations $p_1 = 0$ & $p_2 = 0$ & ... & $p_8 = 0$; the upper index indicates the number of idealizations involved in the formulation of the law):

$$L^{(8)}: \text{if } Cx \text{ \& } p_{1-8}x = 0, \text{ then } Px = l(Vx) \tag{9.1}$$

According to Nowak, the law of value in this formulation "is a very abstract economic statement" and "is preserved in this abstract form up to Chapter X in Vol. III of Marx's *Capital*" (1980, 10). Here according to Nowak, Marx removes the idealization $p_8 = 0$, that is, it now holds $p_8 \neq 0$. Due to this, the equation in (1) changes so that price of x now is a function of the market value of x, the function being now written as $l(Vx + a(p_8x))$ or $l(Vx + ax)$ for short.[128] The law of value thus acquires the following form:

$$L^{(7)}: \text{if } Cx \text{ \& } p_{1-7}x = 0 \text{ \& } p_8x \neq 0, \text{ then } Px = l(Vx + ax) \tag{9.2}$$

According to Nowak, Marx in yet another step considers the case that "supply and demand regulate the market price, actually the deviations of the market prices from the market value" (1992, 256). This thus should stand for the abolition of the idealization that the difference between supply and demand for x equals zero; now it holds $p_7 \neq 0$ and the law of value acquires the following form ("bx" stands for the $b(p_7x)$; $p_{8,7} \neq 0$ for the conjunction $p_8 \neq 0$ & $p_7 \neq 0$):[129]

$$L^{(6)}: \text{if } Cx \text{ \& } p_{1-6}x = 0 \text{ \& } p_{8,7}x \neq 0, \text{ then } Px = l(Vx + ax + bx) \tag{9.3}$$

which should stand for the law of market value under the considerations of demand and supply.

Yet another modification of the law of value should take place when Marx removes the idealization $p_6 = 0$. By this the competition between different branches of production can be taken into consideration; the price of x then does not correspond any more to its market value, but to its price of production, which can be written as $P = k + kp'$. The law of value thus acquires the following form:

---

128 "$a(p_8x)$" expresses the impact of the condition p_8 on the function $l(V)$.
129 "$b(p_7x)$" expresses the impact of the condition p_7 on the function $l(V)$.

$L^{(5)}$: if Cx & $p_{1-5}x = 0$ & $p_{8,7,6}x \neq 0$, then $Px = kx + kxp'x$ \hfill (9.4)

Then, by abolishing the idealization $p_5 = 0$ that merchant profit on x equals zero, one derives the law of production price under the consideration of the merchant profit, *mer* for short:

$L^{(4)}$: if Cx & $p_{1-4}x = 0$ & $p_{8,7,6,5}x \neq 0$, then $Px = kx + kxp'x + merx$ \hfill (9.5)

Finally, Nowak considers the abolishment of the idealization $p_4 = 0$, where then obtains the law describing the price formation of commodities in agriculture which involves already the category of rent; R for short. The law of value then acquires the form:

$L^{(3)}$: if Cx & $p_{1-3}x = 0$ & $p_{8,7,6,5,4}x \neq 0$, then $Px = kx + kxp'x + merx + Rx$ \hfill (9.6)

So, Nowak, like Robinson, interprets the structure of Volume I of *Capital* retroactively from the point of view of the network of categories of political economy derived by Marx only in (the manuscripts of) Vol. III. Thus, for example, if Marx deals in the latter with the effects of the difference between the organic compositions of individual capitals in one branch or between capitals in different branches in an economy, then these differences should be given already in Volume I, where Marx – *it is claimed by Nowak and Robinson* – puts them equal to zero.

Stated otherwise: both Nowak's and Robinson's understanding of the sequence of derivations of categories accomplished by Marx is based on the knowledge of all the categories introduced by Marx in those manuscripts which were intended by him for Book III. Thus, neither Nowak nor Robinson takes into account the fact that Marx's method of theory construction as given in Volume I and in the manuscripts of Book II and Book III involves a *gradual construction of a branching network of categories of political economy, so that at each level of this construction new categories political economy are introduced which were not given at previous levels*.

The same holds for structure of Volume I. At its very beginning categories like surplus value, measure of surplus value, constant capital, and variable capital *do not exist here because they have not yet been derived*. This fact is emphasized by Marx at this very beginning with respect to the category of wage as follows: "The reader has to note that one does not speak here about *wage* or value the worker receives for, say, one day's labor … The category of labor-wage (*Arbeitslohn*) as yet does not exist at all at this level of our presentation" (1987, 78).

In order to overcome the distorted views on the categorial structure and methods of *Capital* of Robinson and Nowak it is necessary to clarify the methods employed by Marx in Volume I and in the manuscripts of Book II and Book III.

In some of these methods Marx drew heavily on Hegel's *Science of Logic* (1986). Accordingly, I have first to clarify the methodological implications of this work for Marx.

9.3 The Methodological Implications of Hegel's *Science of Logic* for Marx

In order to understand the methodological implications of Hegel's *Science of Logic* for the construction of Volume I of *Capital* and of the manuscripts of Book II and Book III, one has to understand, first, the difference between Hegel's and Marx's understandings of what the network of philosophical categories in the *Science of Logic* as a whole stands for.

For Hegel that network should stand for a philosophical "*presentation of God as he is in his eternal essence before the creation of nature and of a finite mind (Geist)*" (1986, Bd. I, 44). In order to understand what it could have stood for in Marx's thought, one should note his reinterpretation – from the point of view of a *realistic epistemology* – of the method of the ascent from the abstract to the concrete, applied by Hegel in the construction of the system of categories in the *Science of Logic*. As Marx states in the manuscript on the method of political economy, this method is "the method to appropriate the [real] concrete, to reproduce it as a concrete in mind" (1976, 36).

His allegiance to Hegel, under such interpretation, stated also in a manuscript to Book II, is as follows. "My relation to Hegel's dialectics is very simple. Hegel is my teacher ... However, I have taken the liberty to behave to my teacher critically, to strip his dialectics of its mysticism, and thereby change it essentially" (2008, MS II, 32).

So, Marx could view the network of categories given in the part "Objective Logic" of the *Science of Logic* as a philosophical reconstruction *of those forms of thinking (Denkformen) by means of which humans in their mind reproduce the real (natural and/or social) world outside their mind*, while the network of categories given in the part "Subjective Logic" he could view as a philosophical reconstruction *of those forms of thinking by means of which humans in their mind create projects for future practical interventions in the real (natural and/or social) world outside their mind*.[130]

Marx's interpretation of the categories unified by Hegel under the heading "Objective Logic" and "Subjective Logic" could thus also be accomplished in

---

130 Marx labeled the creation of such projects by the German term "praktische-geistige Aneignung" of the world (1976, 37).

the framework of a *realistic epistemology*. Under this realistic interpretation, the particular philosophical categories as well as their network as given in the "Objective Logic," would acquire for Marx a crucial *normative* function.[131] Any scientific theory with its specific concepts (for example, political economy with its categories like value, price, etc.) should be constructed in such a way that these concepts and their linking into a network should correspond to those categories and their linking into a network in sense that the latter have, with respect to the former the status of their *general types*.[132]

From the categories of the "Objective Logic" as central for Marx's *Capital* appears the trio of categories, each component of which in turn is viewed by Hegel as a cluster of categories, *appearance – essence (ground) – manifestation*.[133] Marx takes this trio over from Hegel and applies it both for the construction of the first volume of *Capital*, as well as in his planned construction of the entire work.

The category of appearance stands for the knowledge of surface phenomena and phenomenal laws and provides the starting point for the movement of cognition to the essence (ground) of these phenomena and their laws. In this movement that knowledge reappears as re-derived and explained on the basis of knowledge of the ground. This latter knowledge is then given as that of phenomena as manifestations and as laws of manifestation of the ground. The principal difference between appearances and their laws, on the one hand, and manifestations and their laws, on the other hand, is that due to that derivation by explanation *the concepts standing for the knowledge of the essence (ground) of the phenomena and their laws reappear in the conceptual expression of phenomena and their laws as manifestations and their laws*. The cyclical movement *appearance → essence (ground) → manifestation* is succinctly characterized by Hegel by the catch-phrase "retreat into the ground and coming out of it" (1986, Bd. II, 103).

To dispel possible reservations about Hegel's view on the cyclical nature of the movement of cognition which follows the triple pattern of the categories *appearance → essence (ground) → manifestation* it is worth mentioning here that C. G. Hempel (of course, in his *post*-empiricist phase) discerned a method corresponding to that trio when dealing with the issue of theory construction. He states more than 150 years after Hegel the following (1970, 142):

---

131 For a detailed analysis of this appropriation in this framework see (Hanzel 1999, 190–192).
132 On this see (Hanzel and Černík and Viceník 1994).
133 A detailed analysis of this trio of categories is given in (Hanzel 2010) and (Hanzel 2014).

Theories are normally constructed only when prior research in a given field has yielded a body of knowledge that includes empirical generalizations or putative laws concerning phenomena under study. A theory then aims at providing a deeper understanding by construing those phenomena as manifestations of certain underlying processes governed by laws which account for uniformities previously studied.

Which are the central features of the method of ascent from the abstract to the concrete? Here I can draw on Hegel's reflections on the nature of this method as given at the very end of the *Science of Logic*. They can be summarized in the following five-fold way.

First, this method has a certain starting point (1986, Bd. II, 553)

> The beginning ... an *immediate* ... *first of all* ... is not an immediate of *sensual intuition* or of *representation*, but of *thinking* which, because of its immediacy can be termed a suprasensuous, *inner intuition* ... Cognition (*Erkenntnis*) ... is a conceptualizing (*begreifendes*) cognition, therefore its beginning only in the *medium of thinking* – a *simple* and a *universal*.

Second, it holds (1986, Bd. II, 554–555):

> The authentication of the *determined content*, with which the beginning is made ... is to be viewed as going forward, if it belongs to the conceptualizing cognition. The beginning has, therefore, for the method no other determination than that of being the simple and universal; this is precisely the *determinateness*, due to which it is deficient.

Third, (1986, Bd. II, 569):

> Cognition rolls onwards from content to content. First, this forward movement determines itself so that it begins with simple determinacies, and the following become more *rich* and *concrete*. For the result contains its beginning and its course has enriched it with a new determinateness ... The advancement is, therefore, not to be taken as a *flowing* from *other* to *other*. In the absolute method the concept ... raises at each stage of further determination the entire mass of its preceding content and not only does not lose anything by its dialectical advance, or leaves something behind, but carries with it all it has gained, and enriches and densifies itself.

Fourth, one faces here an *extension* (1986, Bd. II, 569) of knowledge due to cognition in which (1986, Bd. II, 570)

> each step of the *advancement* in the further determination, by moving further away itself from the indeterminate beginning, is also a *getting back nearer* to it; and that therefore, that what initially may appear as different, the *retrogressive grounding* of the beginning and the *forward-going further determining* of it, coincide and are the same.

Fifth, this means that "[t]he method ... loops here itself into a circle ... By virtue of the indicated nature of the method, the science presents itself as an in itself looped *circle*, into whose beginning ... the end is being looped back by the mediation; thus, this circle is a *circle of circles*" (1986, Bd, II, 570–571).

As I will try to show below, all these five features of the method of the ascent from the abstract to the concrete play a central role in Marx's overall construction of *Capital*, envisaged by him as a theoretical system given by the intertwining of Book I (the production process of capital), Book II (the circulation process of capital) and Book III (the process of capitalist production as a whole).

The above given interpretation of the cluster of categories unified under the heading of "Objective Logic" and "Subjective Logic" in the *Science of Logic* is also central for understanding of what Marx meant by *categories of political economy*. He understood them as *forms of thinking shared by social actors practically participating in economic processes by means of which they in mind reflect their own practice and create projects of this practice*. These forms can be viewed as *first-level categories* on which political economy as science builds *second-level categories*.[134]

As to the first-level categories, Marx views them as having their origin in "everyday life" (*Alltagsleben*) (1987, 500) and characterizes them as "socially valid, thus objective forms of thought" (1987, 106) even if, very often, they have the character of inter-subjectively shared fictions, like for example the category of value of labor. These he characterizes as "crazy forms" (1986, 106) and as "imaginary expression," (1987, 500) and provides their additional characterization by means of categories taken from the above given trio appearance – essence (ground) – manifestation: "They are categories for the forms of appearance of essential relations" (1987, 500) and then adds (1987, 504):

> By the way, what is valid for forms of appearance – 'value and price of labor' or 'labor-wage' – in contradistinction to essential relation which manifests itself – the value and price of labor power – is valid also for all forms of appearance and their hidden background. The first reproduce themselves directly, spontaneously, as commonplace forms of thinking, the other has as yet to be discovered by science.

According to Marx, what political economy as science produces are second-level categories and there are three possible relations between these categories and their first-level counterparts. One, when second-level categories are just simple restatements and summarizing of the first-level one (this is done by political economy which he labels as "vulgar" one) or, second, when the second-level ones go at least beyond the first-level ones by grasping the essential relations hidden behind the appearances expressed by the first-level ones. The third possible relation, which he views as the *ultimo ratio*, is the accomplishment of a re-derivation and re-statement of the first-level categories by explaining them on the basis of the second-level categories grasping already the essential relations. It is only here

---

134   Here my terminology draws on (Schutz 1954).

that one can, according to Marx, explain why the first-level categories are at all produced in the mind of the actors involved in economic processes.

9.4 Marx's Methods in Volume I and in the Manuscripts of Books II and III of Capital

9.4.1 Capital Volume I

Marx starts by delineating the initial universe of discourse of the theory he is going to construct as that of a *mass of commodities*; "The wealth of nations in which the capitalist mode of production dominates appears as an 'immense collection of commodities'" (1989, 69). And, then, he picks out an individual commodity and subjects it to an analysis. Thus, already at the very beginning, Marx by means of this delineation in fact imposes on his theory construction the requirement to provide at one of its *later* stages a *backward justification* of why he starts at all this construction by choosing the *mass of commodities* as the universe of discourse of this theory. This, as I will show, he accomplishes at the very end of the first volume of *Capital*.

In the individual commodity he distinguishes between its use value and exchange value, the latter being the ratio in which it exchanges against other individual commodities. Then, by means of the exchange value as such shared by all individual commodities, he identifies a quality shared by all of them, namely, that they are the result of abstract labor being expended on their production. And, once they share this quality, then they share also the quality of having *value*.

Because the category of exchange value is for Marx the starting point for the movement to the category of value, he again – and now explicitly – imposes on his theory construction the requirement to re-derive backwardly at a subsequent stage of this construction the category of exchange value. Accordingly, he declares the following. "The progress of the investigation will lead us back to the exchange value as the necessary form of expression or form of manifestation of value" (1987, 72).

Marx also assigns to the category of abstract labor a quantitative characteristic, namely, its time duration, thus assigning to this category a *magnitude* understood as a unity of *qualitative* and *quantitative* characteristics, the former drawing on the category of value.[135] As a consequence, since he views value as

---

135 Here I draw on (Berka 1983), where magnitudes are viewed as unifying the knowledge about qualitative and quantitative characteristics of the entity under investigation; one then speaks about the "size of a magnitude." This terminology differs from

the result of expending abstract labor, he can quantify value by quantifying this labor, the former being determined "by quantum of labor expended during its production" (1987, 72). Thus, he assigns also to the *category* of value its corresponding *magnitude*.

Next, Marx subjects both the categories and magnitudes of value and labor to a combined qualitative and quantitative analysis. The category of labor he characterizes as "identical human labor, expenditure of the same human labor power. The total labor power of the society, which represents itself in the value of the world of commodities, counts here as one and the same human labor power, even though it consists of countless individual labor powers" (1987, 73).

And, at the same time, he characterizes it quantitatively by employing the mathematical term "average" (1987, 73):

> Each of these individual labor powers is the same human labor power as any other, as far as it has the character of social average-labor-power and acts as such a social average-labor-power, that is, requires in the production of a commodity only an average necessary or socially necessary labor time.

This, in turn, leads to a more determinate characterization of individual commodity, initially picked out by Marx from the "mass of commodities," so that now it turns into an "average exemplar of its kind" (1987, 73). With respect to this average exemplar, the magnitude of value acquires a new meaning, namely, that is the necessary labor time required to (re)produce – under the given social conditions of production – a new exemplar of commodity of this kind.[136]

By bringing in additional quantitative characterization of the categories labor and commodity, Marx again imposes on his theory construction the requirement to explain *backwardly* at a *subsequent* stage of this construction how that average is constituted in the economic processes. To be more precise, it requires to explain how labor power and means of production are distributed in an economy among its different industries as well as among particular companies in the particular industries; only under a certain distribution of this elements of production one can characterize an economy by means of average sizes of economic magnitudes. As I will show, Marx does this in the manuscripts of Book III.

By providing a qualitative as well as quantitative characteristic of value by means of the magnitude of socially necessary abstract labor, Marx can, in turn, provide the conditions for the qualitative and quantitative identity of commodities. Two

---

the more common terminological differentiation between "quantity" and "magnitude," the latter referring to the values a quantity actually acquires.

136   On this see (Marx 1980a, 111).

products share the same (social) quality – *being a commodity* – if both are the products of private, mutually independent labors (1987, 75). And once two products *are* commodities, then they are *quantitatively* identical if they are produced by the expenditure of the same quantity of labor (1987, 73).

In a next step, Marx moves to yet another qualitative characteristic of labor, namely, that it is a unity of *concrete* (use value creating) labor and of *abstract* (value creating) labor; thus he views labor as an internally differentiated entity. And, just as a commodity's value is the latter's result, so is value also an inherently differentiated entity. Based on commodities' inherent internal differentiation, Marx then *returns* to the category of exchange value thus *accomplishing the first cycle of theory construction* given in Volume I of *Capital*. This return he introduces with the following statement: "In fact we started from exchange value or exchange ratio of the commodities, in order to track down the value hidden in it. Now we have to return to this form of manifestation of value" (1987, 81).

By a further development of the category of exchange value into the categories of relative and equivalent form, he then, finally, derives the category of *money* and, by means it, the category of *price*. With respect to the former he declares: "The specific kind of commodity with whose natural form socially growths together the equivalent form, turns into money-commodity, or functions as money" (1987, 100) as well as: "[m]oney as the measure of value is the necessary form of manifestation of the immanent measure of value of the commodities, of the labor-time" (1987, 121). With respect to the latter holds: "The simple relative expression of value of a commodity… in the commodity functioning already as money-commodity … is the price-form" (1987, 101).

Why does Marx at all take the torturous cyclical path from the category of exchange-value via the category of value back to that exchange-value and, on its basis, to those of money and price? In my view, the crucial issue here is that Marx tries to prove that money is the *necessary* product of the mutual exchange of products, once the latter are commodities. The inspiration to perform such a proof goes back to Marx's critique – in the manuscript *Grundrisse* – of Proudhon's and of his disciples' attempts to conceptually reflect on the possibility of creating an economy based on commodity exchange bypassing money as the medium of this exchange, and at the same time being aware of the *practical* failure to organize such an exchange.

This is the background on which Marx's gradual introduction of the particular categories and of their respective magnitudes takes place: starting from that of commodity, use value and exchange value, then of labor and value, then of relative and equivalent forms standing for the further development of exchange value, and finally of money and price.

In addition to this method of gradual introduction of categories of political economy and of their corresponding magnitudes, yet another method is applied by Marx already at the beginning of the first volume of *Capital*. What I mean here is the *method of performing variations of the size of these magnitudes and relating them to variations of the size of other magnitudes*; based on this he then states general laws relating these variations. For example, about the relation of variation of the size of value to the variation of the size of labor time he declares (1987, 74):

> The size of the value of a commodity would remain ... constant, if the labor-time required for its production would remain constant. The latter, however, changes with each change in the productivity of labor ... In general: The greater the productivity of labor, the less the labour time required for the production of an article, the less the mass of labor crystallized in it, the less its value. Inversely, the less the productivity of labor, the greater the necessary labor time to produce an article, and the greater its value. The size of value of a commodity thus changes directly as the quantum and inversely as the productivity of the labor realized in it.

Such general laws Marx also states when relating variations in the size of relative value of commodities with variations in the size of their value; (1987, 87):

> Real changes of the size of value are reflected ... neither unambiguously nor exhaustively in its relative expression or in the size of the relative value. The relative value of a commodity may change, although its value remains constant. Its relative value may remain constant although its value changes and, finally, the simultaneous changes in its value-size (*Werthgrösse*) and in the relative expression of this value-size need not correspond to each other.

In a next step Marx passes from the category of money to that of capital by dealing conceptually with the issue of origin of the increment of an amount of money, $\Delta M = M' - M$, in the circuit $M - C - M'$. He expresses by means of the antinomy "capital cannot arise from circulation and it cannot as well not arise from circulation" (1987, 182) the fact that the previously developed network of categories is not sufficient to explain the origin of that increment.

Once he arrived at this conclusion, Marx had in fact two options to eliminate this antinomy. One was to discard the whole network of categories, together with their corresponding magnitudes, constructed by him in the first chapters of Volume I and, thus, viewing this network not only as not sufficient but also as not necessary for the explanation of the origin of the money increment ΔM. The other option was to view this network, even it even if not sufficient, but still as *necessary* for that explanation.

This second option was taken by Marx, which he explained as follows. "The transformation of money into capital is to be developed on the basis of the laws immanent to the exchange of commodities, so that the exchange of equivalents is the starting point" (1987, 182). This means that by using that network, and here especially the category of value, as the starting point for the introduction of additional categories and their corresponding magnitudes – required for the elimination of that antinomy – this network is *preserved by being shifted to the next level of theory construction.*

At this next level Marx subsumes the category of labor power under that of commodity. Based on this he applies to it the *previously developed* categories of value and use value. For the value of labor power holds that it is "identical to that of any other commodity, [it] is determined by the labor time necessary for the production, and thus also reproduction of this specific commodity" (1987, 186).

Its use value is such that it can create, by being employed in the process of production, a value that is greater than its own value. So, while at the previous level of theory construction, Marx dealt with the process of exchange of commodities based on the *value* they have due to the labor expended on their production, now he deals in addition with the process of *valorization* in the course of applying labor – the latter being before that already purchased for its value in the sphere of circulation – in the process of production. On the relations of these processes he states the following (1987, 208–209):

> If we now compare the process of the creation of value and the process of valorization, then the process of valorization is nothing but a extended, beyond a definite point, process of creation of value. If the latter continues only to the point where the value of the labor paid by the capital is replaced by a new equivalent, then it is a simple process of creation of value. If the process of value creation continues beyond this point, then it becomes a process of valorization ... As unity of the labor process and value creation process, the process of production is the process of production of values, as the unity of labor process and valorization process it is capitalist process of production, capitalistic form of value production.

There are three results of the differentiation between these two processes.

First, it leads to the introduction of the category of *surplus value*. The latter is viewed by Marx as a result of the valorization process and so as this process is just a quantitative excess above the time duration of the reproduction of value of the labor power; this category has in Marx approach also the status of a magnitude.

Second, this differentiation enables to solve the above mentioned antinomy; the creation of the increment ΔM: "takes place in the sphere of circulation and does not take place in it. By the mediation of circulation, because conditioned by the purchase of the labor power in the labor market. Not in circulation, because it only initiates the valorization process which takes place in the sphere of production" (1987, 207).

And, third: "All conditions of the problem are solved, and laws of the commodity exchange are in no way violated," (1987, 207) that is, the supposition that commodities are mutually exchanged at a ratio corresponding to their respective value, still retains its validity at this level of theory construction.

From the point of view of the method applied by Marx when passing from the first level of theory construction with the network of categories and their corresponding magnitudes like exchange value, value, labor, labor time, money to the second level, where the category/magnitude of surplus value is already given, it holds that *all the categories and magnitudes from the former level are shifted to the latter*. In order to deal conceptually, for example, with the issue of purchase of labor power in the sphere of circulation, Marx needs the category of money and in order to deal conceptually with the process of valorization understood as a prolongation of the value-creating process, he needs the category of value, labor, labor time, etc.

This method I view as exemplification of the method of the ascent from the abstract to the concrete. That this is so can be seen when one compares Marx's understanding of the value-creation process at the first level with his understanding, at the second level, of the value-creation process on which is already imposed the valorization process, *from the point of view of the social actors involved in these processes*. At the first level of theory construction Marx speaks just about the "producers" (1987, 78) being involved, while "it remains quite undecided (and is in fact indifferent) from whom these objectified labor originates" (1988a, 33), at the second level it is already, in addition, specified that the producer is the laborer performing the total (necessary plus surplus) labor. Thus, a further specification and condensation of the network at the second level is achieved when compared to the network at the first level.

Now, why does Marx perform such a condensation of that network by passing from its first level to its second level, in the first place? The reason is Marx's acknowledgment of the failure of previous development of political economy as science, including Ricardo's *Principles*, to derive/explain the category profit, one of whose forms is the above mentioned increment ΔM, on the basis of the law of value.

By subsuming this increment under the category of profit, Marx imposes on his theory construction the requirement to provide at one of its *later* stages *a more detailed, in depth differentiation of the category of profit*, that is, to re-derive all the antecedently known forms of profit designated by political economy by categories like industrial profit, rent, interest and merchant profit. This re-derivation Marx accomplishes in the manuscripts of Book III.

By viewing the surplus value as the result of the valorization process which the capital, initially advanced for the purchase of labor power, undergoes, Marx can – in turn – further specify the category of capital by differentiating it into the categories of *constant capital* and *variable capital*. Marx, initially, differentiates between them as expressing conceptually the difference between the *objective* (thing-like, *sachliche*) factors of production and the *subjective* factors, the labor power. The former stand for the means of production involved in the process of production and whose value is always transferred, but never reproduced, in the process of production to the value of commodities being produced. The latter, on the other hand, by being employed in the process of production have their value reproduced. This conceptual differentiation between the categories of constant capital and variable capital is then further developed by bringing in the category/magnitude of surplus value. By taking into account that the employment of variable capital leads also to the creation of this surplus, that is, "[t]his part of capital is continuously transformed from a constant size into a variable one" (1987, 219–220), Marx assigns to the category of variable capital its corresponding magnitude. And by taking into account that the value of the constant capital remains partially fixed in the means of production and partially passes to the commodities, but its overall sum remains the same, he assigns also to the category of constant capital its corresponding magnitude.

Once Marx has introduced the category/magnitude of surplus value as well as the categories of constant capital and variable capital together with their corresponding magnitudes, he can further develop the quantitative characteristic of that category/magnitude by introducing the magnitude which he labels as the *rate of surplus value*. And this introduction he accomplishes by employing the method of idealization reconstructed by Nowak, labeled, however, by Marx as *abstraction*.

First, he reflects on the composition of the value structure of a commodity coming out of the production process. This value is given as the sum of the constant capital, c, variable capital, v, and the surplus value, s; that is $c + v + s$; if C is the value of total advanced capital $c + v$, we have for this value $C + s$.

However, for the constant part of the advanced capital it holds that only part of its value is passed to the commodity output, while the remaining part is still given in the means of production. But so as the latter "does not play any part in the process of value creation, one abstracts from it here ... Under the constant capital advanced for the production of value we understand therefore only the value of means of production consumed in the production" (1987, 222). Thus, he puts the value of constant capital which still remains bound to the means of production equal to zero.

But even with this idealization, Marx still views the process of the change of value in the course of the production process as being "obscured by the circumstance that in the consequence of increase of the variable component, the advanced total capital increases as well" (1987, 223). In order to eliminate this obscurity, "[t]he pure analyses of the process requires ... to abstract completely from that part of value of the product, in which only constant capital value reappears, that is, to put the constant capital c = 0" (1987, 223). And then he gives yet another justification for this idealization (1987, 224):

> Insofar the creation of value and change of value are considered for themselves, that is, pure, then the means of production provide ... only the material, in which the fluid, value-creating power should fix itself ... the value of the material is indifferent. It only has to be given in sufficient mass in order to absorb the quantum of labor expended during the production process. Given this mass, its value may rise, or fall, or it may be valueless like land and sea; the process of creation of value and of change of value will not be affected by this. We thus put, for the time being, the constant capital equal to zero.

After applying these two idealization steps, Marx can, finally, move to the quantitative treatment of the valorization of the advanced variable capital by measuring it vis-à-vis itself, that is, as the ratio s/v labeled by Marx as the *rate of surplus value*, s' and which should express the *rate of the exploitation* of the labor power purchased by the variable capital v.

Marx also employs the method of idealization in the course of formulation of general laws relating the variations in the size of a magnitude of political economy with the variations in the size of another magnitude. Here I mention only the formulation of laws pertaining to the relation between the mass of surplus value and the rate of surplus value.[137]

He starts by introducing the idealization that with respect to an individual laborer, "the value of the labor power ... is presupposed as a given, constant magnitude" (1987, 303), that is, the variation of its size is put equal to zero, in symbols $\Delta v = 0$, and then he introduces the magnitude of the total advanced variable

---

137 A much more complicated case of formulation of laws is given in Marx's theoretical treatment of intertwining of the process of production of relative surplus value with that of absolute surplus value (1987, 485–494). Here he formulates general laws, whose validity he views as conditioned by the variation of three factors: the length of the day during which labor is performed (the extensive size of labor), the normal intensity of labor (its intensive size, so that a certain quantum of labor is expended in a certain time), the productivity of labor (a given quantum of labor produces a certain quantum of products). In order to state these laws he considers the following possible combinations of the action of the variation of these factors: "one of the factors constant and two variable," that is, one idealization is given, or "two factors constant and one variable," that is, two idealizations are given or, "finally, all three simultaneously variable", that is, no idealization is given (1987, 485).

capital, V, standing for the value of all labor power simultaneously employed in the process of production, so that, if n stands for the total number of labor powers employed, it holds $V = n \times v$. And given the idealization $\Delta v = 0$, it holds as well that "the value or size (*Umfang*) of the variable capital varies directly with the mass of the appropriate labor power or the number of the simultaneously employed laborers" (1987, 303).

Next he states the following equation. "The mass of the produced surplus value is equal to the surplus value provided by the labor-day of the individual laborer multiplied with the number of the applied laborers" (1987, 304), that is, $S = s \times n$. But so as for the rate of surplus value holds $s' = s/v$, it holds as well $s' \times v = s$ and, thus, it holds $S = s' \times v \times n = V \times s'$. The latter equation states the following. "The mass of the produced surplus value is equal to the size of the advanced variable capital multiplied with the rate of surplus value" (1987, 304).

This equation together with the idealization $\Delta v = 0$ could be viewed as a general law based on which on can perform variations of the size of the magnitudes in involved in it. So, for example, if $\Delta S = 0$, then "the decrease in one factor can be compensated by the increase in another factor. If the variable capital is diminished, and at the same time the rate of surplus value is increased in the same ratio, the mass of surplus value remains unaltered" (1987, 304). And it holds as well that "a decrease in the rate of surplus value leaves the mass of surplus value unaltered, if the size of the variable capital or the number of employed laborers increases proportionally" (1987, 304).

Based on the law $S = V \times s'$, Marx states also the following law: "The rate of surplus value (i.e., the rate of exploitation) and the value of the labor power (i.e., the size of the necessary labor time) being given, it is self-evident that the greater the variable capital, the greater the mass of the produced value and surplus value" (1987, 305). And then gives this law the following interpretation (1987, 306):

> The masses of value and surplus value produced by different capitals, under given value and the same rate of exploitation of the labor power, behave directly proportionally as the sizes of the variable components of these capitals, that is, of their components which have been turned into living labor power.

But this interpretation, Marx emphasizes, "contradicts all experience based on evidence (*Augenschein*)" (1987, 306), namely, that in the case when we have two capitals, where for the first holds $v_1 \gg c_1$ and for the second $v_2 \ll c_2$ – with respect to every total 100 units of each of them – *the former does not appropriate a larger amount of profit than the latter* even if, according to that law, it should.

The existence of the contradiction between that evidence, on the one hand, and the network of categories, on the other hand, and their corresponding magnitudes constructed till now (in the framework of which that law was stated derived) is for Marx an indication of the fact that this network is not sufficient for the elimination of this contradiction. However, he views this network still as necessary for the elimination of that contradiction. This network has to be enlarged by integrating into it additional categories and magnitudes. The latter will fulfill the function of "intermediate links" (1987, 306) mediating between the knowledge given by the as yet non-enlarged network and knowledge expressing that evidence. Marx tried, as I will show below, to bring in these additional categories and magnitudes in the manuscripts of Book III.

Let us now turn to the very end of the first volume of *Capital*. In its first edition one encounters, right after subchapter 6.3 on the modern colonization theory, the following claim (1983, 619):[138]

> Finally, we have for one moment to pick up again the thread where we have dropped it when passing to the consideration of the accumulation ... The immediate result of the capitalist production is *commodity* ... We are thus thrown back to our point of departure, the commodity.

This, and especially the phrase "the immediate result of capitalist production," makes it possible for one to trace the source back to Marx's manuscript entitled "Chapter Six. *Results of the immediate process of production*" (1988a).

From the point of view of this chapter it is worth noting that this manuscript leaves no doubt as to the *cyclical method* applied by Marx in deliberate and conscious manner in the construction of Volume 1 of *Capital*. He characterizes it here explicitly as "the circular course of our presentation" (1988a, 24), and by applying it he justifies at the end of this volume its starting point expressed by the statement: "The wealth of societies, in which the capitalist mode of production dominates appears as an 'immense collection of commodities'" (1987, 69).

The following claims in the manuscript of Chapter 6 made by Marx with respect to that cyclical method are worth to be quoted here in full length (1988a, 24, 27, 30):

> The *commodity*, as the elementary form of the bourgeois wealth, was our point of departure, the presupposition for the emergence of capital. On the other hand, the *commodities* appear now as the *product of the capital* ... If we ... consider the societies of *developed capitalist production*, then the commodity appears in them both as the

---

138 The whole paragraph at the end of subchapter 6.3 was eliminated by Marx from the second edition of *Capital* (1987) and does not appear in the French translation (1989).

constant elementary presupposition of the capital as well as, on the other hand, the immediate result of the capitalist production process ... We start from the commodity, from this specifically social form of product – as the basis und presupposition of the capitalist production. We take the individual product into our hand and analyze the determinations of form which it contains as the commodity, which make of it a commodity ... We treat the commodity as such a presupposition by starting from it as the simplest element of the capitalist production. On the other hand, however, the commodity is a product, result of the capitalist production. What is first its element, presents itself later as its own product.

Given the re-derivation of the starting point of the first volume of *Capital*, the universe of application of the category of value and of its corresponding magnitude undergoes a transformation. Now, both of them can be applied only to the mass of commodities produced by the economy in question. At the same time, what remains to be discovered is the characteristic of the capital employed in the production of the average exemplar of a kind of commodity from this overall mass of commodities.

Marx also clearly expresses the yield of his cyclical approach to the construction of Volume I, namely, that it leads to an *extension* of knowledge; (1988a, 33):

> The *commodity*, as it comes out from the capitalist production, differs in its determination from the commodity we started from as the element, presupposition of the capitalist production. We started from the individual commodity as an independent article in which a certain quantum of labor time has been objectified and which, therefore, has an exchange value of a certain size. The commodity appears now as further determined in a twofold way (*doppelt bestimmt*): ... 1) the *commodity* as *product of the capital* contains partially paid, partially unpaid labor ... 2) The individual commodity appears not only materially as a part of the total product of capital, as an aliquot part of the total (*lot*) produced by it, We no longer have before us the individual independent commodity, the individual product. Not individual commodities appear as the result of the process, but a *mass of commodities*, in which the value of the advanced capital + the surplus value – the appropriated surplus labor – has been reproduced.

This extension of knowledge is used by Marx at the very end of the first edition of *Capital* in order to provide a link to Book II. By moving to it, he builds on the view he presented when moving from the category of money to that of capital, namely, that the latter is a unity of the processes of production and circulation. And, once the commodity is understood already as a result of the process of production and at the same time as an embodiment of a certain amount of surplus value, then "[w]e are thrown back to our point of departure, the commodity, and with it to the sphere of circulation. But what we have to consider in the following book is no longer the *simple commodity circulation*, but *the circulation process of capital*" (1983, 619).

9.4.2 Manuscripts of Book II

Marx starts several manuscripts intended by him for Book II of *Capital* with the statements which are identical with those given at the very end of the first edition of Volume I. He states: "We have seen that the total product of the capital presents itself a mass of commodities" (1988b, MS I, 141) and in another manuscript he declares in a similar manner: "The immediate result of the capitalist process of production is a mass of commodities" (2012, MS IV, 286).[139]

The overall aim of this Book he then characterizes with respect to categories of political economy as follows: "to develop the pure *form-determinations* (categories), the creation of new form-determinations of the capital consequent upon its passage through the process of circulation" (1988b, MS I, 141).

In order to perform this development he gradually imposes the following idealizations:

1) the components involved in the reproduction processes have a constant value, that is, the productive power of labor producing them does not change (1988b, MS I, 141).
2) The commodities are sold and purchased at prices corresponding to their respective values; "We disregard the fluctuations of the market prices" (2008, MS II, 37).
3) "In order to simplify the presentation we start initially from the supposition that the whole advanced capital now exists in the form of salable commodities or, what is the same, we abstract from that part of capital which does not reproduce itself in this form" (1988b, MS I, 144).
4) It is presupposed that all moments of the capitalist production process are results of the capitalist type production process and not of non-capitalist types (1988b, MS I, 141), that is, it is "the *capitalist type of production* is not only the dominating but the universal and exclusive form of production" (1988b, MS I, 306).

Based on these idealizations, Marx investigates into the circular courses of money, commodity and production capitals. Next, he introduces the category of circulation time (*Umlaufszeit*) of capital which has the status of a magnitude referring to the time in which the capital is bound to the sphere of circulation (2008, MS II, 53) and then he introduced the category of turnover time (*Umschlagszeit*)

---

139   For a better orientation I add in the references also the number of the respective manuscript (MS).

given as the sum of the production time and circulation time. The former magnitude enables to state the following quantitative relations (2008, MS II, 53):

> The more *labor time* (*Arbeitszeit*) a capital of a given size of value can make liquid, the larger its self-valorization. The larger, on the contrary, the circulation time of a capital with a given size of value, the smaller its self-valorization. The more the form-transformations in the sphere of circulation are only ideal, that is, its time duration approaches zero, the more the productive function, given a certain rate of surplus value, reaches its *maximum*.

Worth noting is that in the manuscripts of Book II Marx puts – as compared to Volume I – a *stronger emphasize on quantitative treatment of economic processes* and where he draws on the already derived magnitudes. This is readily seen in the following two cases.

First, he investigates how the time of the production process influences the overall turnover time of capital. In order to grasp this influence he compares two capitals employed in two different kinds of production processes (cotton mill and production of steam locomotives) which differ by their respective time of production.

Their comparison is performed by Marx under the following six idealizations: (i) the difference between the respective size of these capitals is zero; (ii) the difference in their respective division into constant and variable parts of is zero; (iii) the difference in their respective division into circulating and fixed parts of capital is zero; (iv) the difference in their respective labor days is zero; (v) the difference of the respective division of the labor day into necessary labor and surplus labor is zero; (vi) in both productions the difference between the moment of delivery of the commodity and the moment of payment for it equals zero. Based on these idealizations, Marx states: "An extension of the production time thus extends ... the turnover time or decreases the turnover velocity" (2008. MS II, 182).

Second, with respect to the magnitudes related to capital, namely, production time, circulation time and turnover time, Marx considers the impact of changes of the circulation time on the overall turnover time while holding the size of the magnitude of production time constant. He declares: "It goes without saying that a different length of the circulation time affects the turnover time a[nd] thus the size of the turnover period a[nd] differentiates depending on its relative length" (2008, MS II, 208).

In order to prove this, he applies the following set of methods. (2008, MS II, 208–209):

> Most palpably this becomes visible, either if one compares two different capital investments, where all other circumstance which modify the turnover are the *same* (identical) a[nd] only the circulation times are different or, if one takes a given capital with a given

237

composition o[f] fix and fluid capital, given labor period etc., and hypothetically varies only the circulation time.

And, then, the central operation is the cancellation of the idealization that the money is paid at the moment when the commodity is delivered; "the one portion of the circulation time – a[nd] the relatively most decisive – consists of the *selling time* (*Verkaufszeit*), the period when the capital is in the state of commodity capital" (2008, MS II, 209) and which involves also the time required for the transportation of the commodity.

Finally, let us turn to Marx's treatment of the process which he labels in one of the manuscripts of Book II as "The real conditions of the circulation a[nd] reproduction process" (2008, MS II, 340). He labels by this the interlinking and intertwining of individual capitals to the total social capital. From the point of view of this chapter at least the following two aspects of Marx's treatment of this interlinking and intertwining should be mentioned here.

First, Marx tries to understand here the *quantitative* proportions in which the two groups of production – of products for personal consumption and means of production – exchange against each other so that the reproduction of total social capital can take place at all. And his treatment of these proportions is performed under the idealization that prices at which commodities are mutually exchanged are always equal to their respective values. Given this idealization he then considers, first, the idealized case of simple reproduction, that is, when that part of surplus value which turns into capital is now put equal to zero. And, then, he abolishes this idealization and considers the case of enlarged reproduction at the level of the total social capital.

Second, Marx employs a cyclical method in the manuscripts of Book II. Initially, he presents a characterization of capital which is fulfilled by any and all individual capitals, while supposing that it is being shaped and modified by its interaction – by means of exchange – with the remaining, total social capital. But how this shaping and modifying takes place can be explained only when dealing with the mutual interactions of individual capitals making up the total social capital. Thus, Marx's initial explanation of the circuits of any and all individual capitals involves unexplained moments which are explained only *backwardly* by *advancing from this explanation* to the explanation of the functioning of the total social capital. This cyclical method of explanation which Marx planned to apply in Book II he delineates as follows (2008, MS V, 636–637):

> That the social capital = sum of individual capitals ... a[nd] the total movement of the social capital = sum of the movements of individual capital, does not preclude in any way that this movement as movement of the isolated (*vereinzelt*) individual capital

presents other phenomena as the same movement when it is considered from the point of view of a partial movement, as a motion of a part of the total motion of the social capital, that is, in its connection with the movements of its other parts, a[nd] that at the same time this solves problems whose solution is *presupposed* when considering the circuit of the isolated individual capital, instead of being the result of it.

9.4.3 Manuscripts of Book III

In the manuscripts intended for Book III, Marx continued to employ the method of theory construction based on the derivation of different form determinations of capital. In the main manuscript of this book he states (1992, 7):

> We have seen that the process of production considered as a whole is a unity of the process of production and circulation ... What is dealt with in this book cannot be to perform general reflections about this "unity." It is rather to find and present the concrete forms which grow out from the process of capital concerned as whole.

And those successive derivations are accomplished by the introduction of new categories of political economy, of their corresponding magnitudes and of laws relating them.

The point of departure are the categories of *profit*, p, and *rate of profit*, p'. The former is based on the categories of surplus value and total capital $(C = c + v)$, while the latter on the ratio of the size of surplus value with the size of the total capital "in order to compute its relative size" (1992, 51), that is, s/C. And so as s/v, the rate of surplus value, and s/C yield different sizes, Marx identifies behind this difference the existence of a magnitude different from the magnitude of measure of surplus value; the magnitude of the rate of profit which "as a special category is rightly fixed in a special name" (1992, 51).

The justification of the introduction of the term "rate of profit" by means of the chain *different sizes* → *different magnitudes* → *different categories* of *political economy* Marx justifies as follows: "They are two different measurement of the same magnitude, which thus express two completely different relations (*Verhältnisse oder Beziehungen*), since the difference of the scales (*Maßstäbe*) includes the difference of the relation and ratio (*Verhältnis und Beziehung*)" (1992, 52).[140]

---

[140] My translation of "Verhältnisse oder Beziehungen" as "relations" is based on Marx's employment of the German connective "oder" which should unify words he views as synonymous. My characterization of Marx's introduction of the term "rate of profit" by means of this chain is based on his employment of the German term "Grösse," which has two different meanings which in (Berka 1983) are expressed either by the term "size" or by the term "magnitude." This meaning ambiguity as given in German leads to the linguistic pun "die Grösse einer Grösse." By differentiating in

Marx provides yet another justification of that introductory chain going from the category of rate of surplus value to that of rate of profit by drawing on categories taken from the above mentioned trio appearance – essence (ground) – manifestation (1992, 52):

> The transformation of surplus value into profit is to be derived ... from the transformation of s' into p', not the other way round. And in fact, the rate of profit is that from which one historically is starting, surplus value and rate of surplus value are, relatively, the invisible and the essential to be investigated, while rate of profit and thus form of surplus value as profit show themselves on the surface of the appearances (*Erscheinungen*).

Based on the category/magnitude of the rate of profit Marx, then, states several laws pertaining to the changes of the size of the magnitude of rate of profit taking place under certain idealizations. The methodological norm being here at work he states as follows (2003a, 128):

> When considering the rate of profit we assume ... a given capital with a given composition a[nd] a given rate of valorization. We then run the possible sequence of changes which give rise to changes in the rate of profit, the latter being a function of several variables, a[nd] we find the laws which determine the rise, decline or constancy of the rate of profit, in one word, the laws of its movement.

Among these laws there are several which he labels as the "main laws" (*Hauptgesetze*) for example: "*The rate of profit is to the rate of surplus value as the variable capital is to the total capital. It rises and falls as the proportional size of the variable capital (here presupposed that s' is given)*" (1992, 26–27).[141]

As the next category and its corresponding magnitude Marx introduces the category of *cost-price* of a commodity understood as the value of the capital used up in its production. And, then, given this category/magnitude and by taking into account that a commodity in addition to the capital employed in its production incorporates also a surplus value, states the "law ... that the cost-price of a commodity [is] smaller than its *value*" (1992, 56).

Next, based on the categories of profit, rate of profit and cost-price, Marx derives the categories of *average rate of profit, average profit, production price* and *market value*. These derivations he performs under several idealizations (1992, 231).

First, there exists no difference between the branches of production as to speed with which the constant capital enters into the annual product of the capitals employed in these branches.

(Berka 1983) between these two English terms it is possible to escape it; this German phrase is then translated as "the size of a magnitude."

141 These laws are also stated in (Marx 2012c) and (Marx 2003a).

Second, there exists no difference between the branches of production as to the size of the annual surplus value realized by capitals employed in them in proportion to the size of their respective variable capitals, that is, the "one abstracts from the *difference* which the diversity of the circulation times can produce" (1992, 231).

Based on these idealizations, Marx derives – at least in his view – the category of average rate of profit by considering the competition between capitals employed in the different branches of the economy. From the point of view of this chapter, the following aspects of Marx's *method* derivation employed here are worth to be mentioned.

First, he is well aware of the fact that the whole chain of categories developed by him leads to the conclusion that once in those branches capitals are given whose organic compositions of capital are different, then – even under the given idealizations – then the rates of profit given in them should be different. But he also aware of the fact that "there is no doubt that in reality ... the difference of the *average rates of profit* for the different branches of industry *does not exist*" (1992, 230).

Second, by acknowledging this contradiction between, on the one hand, the result of his theory construction, and, on the other hand, an economic phenomenon, he acknowledges also that what is under thread of a refutation is not only that chain of categories, but also, and primarily, its starting point – the category of value: "It seems thus that the *theory of value* is not reconcilable with the *real movement* (not unifiable with the real phenomena)" (1992, 233).

Third, but – like in the case of the elimination of the contradiction between theory of value and the existence of the production of the money increment ΔM – he still holds to the whole chain of the previously developed categories. And, he declares about his derivation of the category of average rate profit, that it "presupposes the average rates of profit in the particular spheres of production ... and [which] could be developed only from the value of the commodity. Without this development, the general rate of profit ... remains a meaningless representation devoid of any concept" (1992, 223).

Based on the category of average rate of profit he then derives the category of *production price* of a commodity understood as the sum of its cost-price plus the average rate of profit with respect to the capital applied to its production.

While the category of average rate of profit pertains to the economy as a whole, Marx derives also the category of *market value* by taking into account the competition in a branch of the economy. He delineates this category as follows: "The *market value* is to be viewed, on the one hand, as the *average value* of the commodities produced in *one* branch [and], on the other hand, as the individual value of commodities produced under average conditions of the sphere and which form the *great mass* of them" (1992, 253). And, then, he gives an additional characterization of this category by considering this average case, while only a minority of commodities is produced either under better or worse than the average conditions, so that these extremes cancel each other. As a result (1992, 257):

> the *average value* of each aliquot part of the whole mass of commodities = the individual value of commodities produced under average conditions ... In this case, the *market value* or the *social value* of the commodity mass – the necessary labor time contained in it – [is] determined by the value of the large mass at the middle (*mittlere Masse*).

The derivation of the categories of market value and production price enables Marx also, finally, to close two cycles whose starting point is given at the very beginning of the first volume of *Capital*. By explaining the constitution of the market value he can, retroactively, explain how the average exemplar of a commodity of a certain kind is produced at all, namely, that it belongs to the mass of commodities produced in a certain branch under the dominating average conditions.

The derivation of the category of the production price enables to explain, again backwardly, how the distribution of labor power and means of production among the various branches of a capitalist economy – presupposed without any justification at beginning of *Capital* – is at all established. According to Marx "[t]he *price of production* involves the average profit. We gave it the name *price of production* ... because in the long run it is the condition of the supply, of the reproduction of the commodity in each particular sphere of production" (1992, 272). This means that once the actual prices at which commodities are exchanged deviate from their respective production prices, then, the scale of reproduction in the particular branches is changing, which in turn causes a change in the distribution of the labor power and material conditions of production. Stated otherwise: the category of production price enables to explain the mechanism, by the action of which the distribution of social capital between different branches takes place.[142]

9.5 Methodological Conclusions

The above given analysis of *Capital*, of its Volume I and of the manuscripts intended by Marx for its Book II and Book III, shows that that they are constructed by an employment of the following three methods. *First*, a gradual introduction

---

142 On this see also (Marx 1980b, 1603).

of categories which has a cyclical nature and by means of which a gradual extension of knowledge is obtained. *Second*, a gradual introduction of magnitudes which correspond to these categories. And, *third*, formulations of laws stating the relations between the variation of the sizes of these magnitudes, where these laws are valid under certain idealizations standing for the supposition that the variations of some sizes of some other magnitudes equal zero. These three methods in their unity I view as the main components of the method of the ascent from the abstract to the concrete as employed in Marx's *Capital*.

Based on the reconstruction of these three methods, one can make at least the following two methodological arguments in favor of Marx's network of categories, magnitudes and laws.

First, once this network is given, it provides the means to perform variations of sizes of magnitudes which were not performed by Marx and thus to derive new laws not derived by him, as well as to cancel idealizations not cancelled by him and thus derive yet other laws. So, for example, one could make an attempt at a further development of Marx's economic theory by cancelling the idealization, employed by him in the derivation of the category of average rate of profit that the difference between rates of surplus value between different branches is zero.

Second, this network can be modified also when facing new facts not known before. This very often requires the introduction of new categories and their corresponding magnitudes as well as the formulation of new laws for the variation of their size under new, previously not stated idealizations.

This means, from a general methodological point of view, that *Marx's network of categories, magnitudes, idealizations and laws stated under certain idealization conditions is open to further extensions.*

Based on the characterization of the methods employed by Marx in Volume I and the manuscripts of Book II and Book III, the following evaluation of the views of both Robinson and Nowak can be given.

Contrary to the claims made by Robinson, Marx treats the category of *capital* in Volume I as expressing a *social relation given under certain social conditions and characterizes it by gradually constructing a network of categories, of magnitudes corresponding to them and of laws relating the variations of the sizes of these magnitudes under certain idealization*. The issue of the mutual interactions of *particular* capitals as capitals is not treated in this volume at all; thus, for example, neither *the supposition of identity of their respective organic compositions* nor *the supposition of identity of their respective rates of surplus value is given here*. These idealizations are performed only in the manuscripts of Book III.

As to Robinson's claim about the *metaphysical* status of the category of value in Marx, it holds, since this category not only takes center stage in Volume I but is also shifted to the manuscripts of Book II and Book III, that there does not exist a gap – claimed by Robinson – between a metaphysically burdened Volume I and a "scientific" Volume III free of any metaphysics. If one accepts Robinson's claim that the category of value is metaphysical concepts, then he as to view also Volume II (Book II) and Volume III (Book III) as metaphysical as well.

With respect to Nowak, the following can be stated. Given the methods of gradual introduction of categories, of their corresponding magnitudes and of laws relating the variations of the latter's sizes as applied by Marx in Volume I of *Capital*, it becomes readily seen that what Nowak has reconstructed as the procedures of idealization and of their gradual abolishment, can take place in the framework of this volume as well as in the manuscripts of Book II and Book III only *if* and only *after the respective categories have been introduced and the corresponding magnitudes have been assigned to them. Only then one can reflect on the zero and/or non-zero variations of sizes of these magnitudes.*

Because that introduction remained hidden to Nowak, he fell into the illusion of an *a priori* given complete network of categories/magnitudes some of which or their variations are initially put equal to zero and then, in the course of theory construction these idealizations are gradually cancelled. This methodological misunderstanding of the structure of *Capital* let Nowak, then, to the false claim (1980, 10) that the law of value in the form as given in (9.1) is preserved by Marx up to Chapter X in this volume. The truth, however, is that the law of value undergoes already in the framework of the first volume of *Capital* at least the following two profound transformations.

First, at its beginning this law holds for an average exemplar of commodity of a certain kind. This means, to use a formal parlance, that the individual variable x employed by Nowak in (9.1) ranges at the beginning of that volume over a set of in such a manner characterized *individual* commodities. But, at the end of this volume the law of value holds only for a mass of commodities of a certain kind, that is, x ranges here already over a set whose elements are *sets* of commodities.

Second, at the beginning of the Volume I the average exemplar commodity has "just" a value given by the necessary labor embodied in it. At its end, value is further conceptually differentiated by applying to it categories introduced in the course of the construction of this volume. By means of this differentiation its value is now understood as involving two components – one being the result of the employment of necessary in the sense of paid labor and another being the result of surplus in the sense of not paid labor.

Let us now, turn, finally, to the relation between Hegel and Marx, from the point of view of the method of the ascent from the abstract to the concrete employed by the former in the construction of the *Logic of Science* and by the latter in the construction of *Capital*, Volume I and of the manuscripts of Book II and Book III.

While did Marx take this method from Hegel? The answer is, at least in my view, that Marx saw the strength of Hegel's method as proved by the fact that the latter employed it successfully in the construction of network integrating over two hundred philosophical categories. So, at least in my view, Marx could have viewed this method as suitable also for the construction of his network of categories of political economy. This network, according to my first tentative count, integrates at least thirty such categories.

Since Marx applied the cyclical feature of this method, as shown above in the construction of Volume I as well as of Book II and Book III, it may seem that he made an attempt, like Hegel, at the creation of a network of categories as a *purely self-justifying* system. However, as shown above, Marx's network is, due to the methods employed in its construction, open to the theoretical treatment of new economic facts.

That this is so can also be shown when one compares Marx's employment of this method with that of Hegel. The latter characterizes it as follows (1986, Bd. II, 556–557):

> Essential is that the ... method finds and recognizes the *determination* of the universal within itself ... The ... method ... does not behave in the manner of external reflection but takes the determinate from its subject matter ... This is what Plato demanded of cognition: *to consider things (Dinge) in and for themselves*; partly in their universality; but partly it should not stray away from them and grasp at circumstances, examples and comparisons ... [it] should have them alone before itself and bring to consciousness what is immanent in them. – To this extent the method of absolute cognition is *analytic* ... But it is also *synthetic* as its object, immediately determined as a *simple universal*, shows itself through the determinateness ... as an *other*.

Marx's derivation of categories, so as he gradually derives different form-determinations of capital, shows capital at each level of theory construction as, compared to the previous level, as an "other" and, therefore, yields an *extension of knowledge* about capital. The method he applies is, thus, of *synthetic* nature.

But this derivation is not analytic in the above given Hegelian sense. That this is so can be seen, for example, at two main point breaking points in the first volume *Capital*.

First, when Marx derives the category of money, he does so vis-à-vis the realization of the practical failure of Proudhon's economic theory to organize an exchange of commodities while bypassing the employment of money.

Second, when trying to explain the origin of the increment ΔM by introducing categories like valorization, surplus value, etc., he does so vis-à-vis the realization of the failure of Ricardo's school to link the category of value with that of profit; the latter being the evident result of economic processes under capitalism So, in both this cases, *into Marx's sequence of derivations of categories enter information* – of course, expressed and mediated by means of categories of political economy – *about the state of affairs in the world of economics.*

Chapter 10: Adam Smith's Method of Theory Construction in Book I of *Wealth of Nations*[143]

The aim of this chapter is to reconstruct and analyze the structure of some of the key parts of Book I of Adam Smith's *Wealth of Nations* (henceforth, *WoN*).[144] In contrast with the more common approaches to the analysis of *WoN*,[145] we shall focus on the *methods* employed by Smith in the introduction and derivation of economic concepts, as well as on the *categories* explicitly or implicitly involved.

By the structure of Book I of *WoN*, we mean the set of relations among the key concepts and propositions of Smith's theory. Therefore, the methods we shall focus on in this chapter are of a "conceptual" nature. As opposed to the application of empirical methods, the application of conceptual methods does not result in any new data on particular states of affairs in the world. Instead, conceptual methods yield new concepts or relations among preexisting concepts (e.g., via introduction by definition), or new propositions or relations among propositions (e.g., via the various types of inference). Hence, the application of conceptual methods is a *conditio sine qua non* of theory construction in any scientific field – the social sciences notwithstanding. More precisely, the application of conceptual methods is what the very process of theory construction primarily consists in. And even though the application of conceptual methods does not, by itself, lead to new data, the knowledge obtained by these methods can be used to predict and/or search for new data and to drive their collection.

In our analysis of the conceptual structure of Smith's theory, we shall distinguish two different but closely related levels: the conceptual and the meta-conceptual. The *conceptual* level is that of Smith's economic concepts like value, labor commanded, natural price, profit or rent. These concepts are the basic building blocks, so to speak, of the whole of Smith's conceptual edifice. The *meta-conceptual* or *categorial* level is that of the "meta-concepts" or categories

---

143 This chapter was originally written as a separate paper in cooperation with Dr. J. Halas. Here we drew partially on (Halas 2013).
144 All references are to (Smith 1981). Since we only refer to Book I, we use two digits to indicate the chapters and the respective paragraphs; for example "v. 7" refers to Chapter V, paragraph 7.
145 See, for example, (Kaushil 1973), (Hueckel 2000), (Henry 2000), (Roncaglia 2006), (Dooley 2005), (Pack 2010) and (Naldi 2013).

like cause, measure and standard, which form a general meta-conceptual and methodological framework of *WoN*.[146] This framework provides the more general concepts *in terms of which* Smith presents conceptual knowledge at the first level; it also enables him to perform various operations at that level, via the application of conceptual methods.

Our primary aim is to draw attention to the relatively high level of sophistication achieved in Smith's use of conceptual methods. Specifically, we will focus on the ways in which Smith overcomes certain problems of economic measurement and arrives at a significant growth of knowledge *vis-à-vis* the everyday knowledge of economic agents. Second, we will also point out some of the common errors in interpretation of Smith that we believe are based on misunderstandings, on the part of interpreters, of Smith's employment of categories and conceptual methods. Third, our investigation also has a normative significance, in that we identify certain shortcomings in Smith's approach which weaken the internal coherence of his theory, and could be rectified by a stricter application of conceptual methods.

10.1 Exchangeable Value and Its Measure in Chapter V

At the very end of Chapter IV of *WoN*, Smith formulates his well-known plan of further investigation, framed as an inquiry into "the rules which men naturally observe in exchanging them either for money or for one another ... rules [which] determine what may be called the relative or exchangeable value of goods" (iv. 12). His primary goal is to show "what is the real measure of this exchangeable value; or, wherein consists the real price of all commodities" (iv. 15). Once identified, this "real measure" can, at least in principle, serve as a "standard by which we can compare the values of different commodities at all times and at all places" (v. 17) and be put to use for econometric purposes. In the context of Smith's more general concern with laying bare the causes of development of modern commercial societies, Smith's analysis of concepts like exchangeable value and labor commanded, as well as his employment of categories like measure and standard, serve, in the long run, to link the exchange ratios of commodities and their changes to the methods of production and their changes. Once this link has been established, it enables Smith to move beyond the conceptual space of

---

146 As Kant famously claimed in his first *Critique*, "we cannot *think* any object except through categories" (B165; 1998, 264); for a more recent take on the relationship of categories and cognition, see (Walsh 1953). On categories and their meta-conceptual function with respect to empirical science, see (Hanzel an Černík and Viceník 1994).

the everyday knowledge of agents involved in the process of exchange and production of commodities. In this section, we shall reconstruct Smith's approach to the establishing of this link, drawing partially on O'Donnell's detailed analysis.[147]

At the beginning of Chapter V, Smith defines the extent of one's wealth – or the degree to which one is "rich or poor" – as the "necessaries, conveniences and amusements" (v. 1) one can afford. He implies that the acquisition of such objects is mediated through labor. Thus, in a situation of a developed division of labor, one acquires a great part of these objects through exchange of the products of one's own labor for the products of the labor of others. One's wealth is then determined by the "quantity of labour" of others that one can purchase or command. Smith then defines the (exchangeable) value of a single commodity as the quantity of labor which it enables one to command on the market, i.e., as *labor commanded*. Thus, labor is the "real measure" of the exchangeable value of all commodities.

In the subsequent paragraph, Smith identifies "the real price of every thing" with "the toil and trouble of acquiring it" (v. 2). The meaning of this occurrence of "to acquire" is synonymous with "to produce", as the following lines imply (v.2):[148]

> What is bought with money or with goods is purchased by labour as much as what we acquire by the toil of our own body. That money or those goods indeed save us this toil. They contain the value of a certain quantity of labour which we exchange for what is supposed at the time to contain the value of an equal quantity. Labour was the first price, the original purchase-money that was paid for all things. It was not by gold or by silver, but by labour, that all the wealth of the world was originally purchased; and its value, to those who possess it and who want to exchange it for some new productions, is precisely equal to the quantity of labour which it can enable them to purchase or command.

Two aspects of Smith's considerations need to be emphasized here. First, in addition to *labor commanded*, Smith brings in another quantity, namely, that of (the value of) *labor "contained"* in a commodity – known in secondary literature on Smith as *"labor embodied"*.[149] Second, in the context of pre-capitalist commodity exchange, to which the first three paragraphs of Chapter V apply, the quantities of labor commanded by a commodity and labor embodied in a commodity will be exactly equal.[150]

---

147 See (O'Donnell 1990).
148 For a different interpretation of Smith's "to acquire", see (Naldi 2013).
149 To the best of our knowledge, Smith himself never used the term.
150 The pre-capitalist focus of the first three paragraphs of Chapter V is acknowledged by a host of disparate interpretations (O'Donnell 1990, 64), although not universally accepted – see (Hueckel 2000).

As we have seen, Smith's goal was to link changes in exchange ratios to changes in methods of production, where for capitalistically produced commodities the equality of labor commanded and labor embodied generally does not hold. Knowledge about *changes* in the exchange ratios of commodities could *in principle* serve – in a pre-capitalist and, under some assumptions, also in a capitalist context – as the basis for the quantification of *changes* in the amount of labor expended in the production of these commodities, i.e., changes in the methods of production. However, although perhaps principally possible, the derivation of the latter quantity from the former is not directly possible at the level of *actual measurement* – or econometrics, for short. Smith lists the following reasons for this: "it is often difficult to ascertain the proportion between quantities of labour" due to differences in "hardship endured" and "ingenuity exercised" in the labor process (v. 5); every commodity "is more frequently exchanged for, thereby compared with, other commodities than with labour" (v. 6); and, finally, the medium of exchange of commodities, once barter ceases, is a money-commodity (gold and/or silver), so that while labor commanded is, according to Smith, the "real measure" of exchangeable value, it is not the *standard* actually employed in exchange.

From the point of view of Smith's major aim – that of linking changes in exchange ratios to changes in methods of production – one then faces the following question. Could the change in the price of a commodity, once it is expressed as a variation in the amount of gold or silver for which it is exchanged, by itself be used to measure the variation in the quantity of labor commanded by it and, by proxy, the variation in the quantity of labor expended in its production and thus changes in the methods of production of this commodity? Smith's answer to this question is unambiguously negative, for at least two reasons.

First, according to Smith, the ratios in which precious metals (as money) are exchanged for other commodities change not only due to changes in the methods of production of the latter, but also due to changes in the methods of production of the former. Because the value of precious metals varies, the money commodity "can never be an accurate measure" of value. Smith contrasts this property of gold and silver with labor: "never varying in its own value", it is therefore "alone the ultimate and real standard" of value.[151]

---

151 All quotations in this paragraph refer to (v. 7).

Second, even if one could *in theory* abstract from the changes in the production of precious metals, the variations in exchange ratios of money and commodities still could not be directly linked – at the *econometric* level – to the changes in quantities of labor commanded by a given sum of money and, in turn, to changes in the quantities of labor embodied and thus to changes in the methods of production. The reason is that, according to Smith, the required data are missing: "the current prices of labor at distant times and places can scarce ever be known with any degree of exactness" (v. 22).

As a way out of these problems, and to succeed in establishing the sought-after link (in the context of capitalist exchange) Smith performs two important conceptual shifts.

First, while initially he used the expression "quantity of labor which is commanded by a commodity" as synonymous with the expression "the quantity of (the value of) labor contained in a commodity" (i.e., labor commanded equals labor embodied), he now defines the term as *the quantity of labor that can be purchased by an employer from a laborer*, where this quantity of labor is purchased "with a greater and sometimes with a smaller quantity of goods" (v. 8).

Second, and hand in hand with the first shift, Smith redefines the term "real price of labour". Initially, Smith specifies that: "Equal quantities of labor, at all times and places, may be said to be of equal value to the labourer ... he must always lay down the same portion of his ease, his liberty, and his happiness" (v. 7). The real price of labor is thus equal to the amount of "toil and trouble" expended. Subsequently, in what O'Donnell calls "an important switch in perspective" (O'Donnell 1990, 65), Smith proposes to view real price of labor as "consist[ing] in the quantity of necessaries and conveniences of life which are given for [labor]" (v. 9). In other words, the real price of labor is now defined as the "subsistence of the labourer" (v. 15). Both meaning shifts signify a definite change of focus from pre-capitalist to capitalist exchange.

The two meaning shifts serve to complicate Smith's effort in that they dissociate the quantities of labor commanded by a commodity and the labor "contained" in a commodity. In order to succeed in his long-term aim mentioned above, Smith is now compelled to bring in the assumption that the subsistence of the laborer consists (predominantly, at least) in corn, as well as two additional assumptions: that the physical amount of the corn wage is nearly constant over long periods of time (v. 15) – w_1, for short – and that the cost of corn is nearly constant as well (xi.e.28) – w_2, for short. These will enable Smith to establish a relationship between labor embodied and labor commanded in a situation of capitalist exchange. As we will now show in more precise terms, the two meaning shifts, once unified with these assumptions, enable Smith to perform certain econometric computations.

10.2 Smith's Measure of Value and Econometrics

As shown above, Smith justifies his emphasis on the real (corn) price of labor at the expense of its nominal (money) price by the fact that while the latter prices were not recorded over longer periods of time, the former "are in general better known and have been more frequently taken notice of" (v. 22). In this section, we will clarify how data on corn prices, once interpreted on the basis of Smith's theory, can be employed for econometric purposes.

Smith's theory assumes that (a) quantities of commodities of different kinds contain quantities of labor; that (b) when the former are exchanged, the latter can be said to have been exchanged as well; and that (c) the ratio (i.e., a higher-order quantitative relation) in which the quantities of commodities are exchanged is proportional to the ratio (also a higher-order quantitative relation) of the quantities of labor embodied in them.[152] From this one can infer, even though Smith does not state this explicitly, that (d) the very exchangeability as a qualitative characteristic shared by commodities has its *causal* basis in the existence of labor embodied as such – that is, in labor embodied as their common quality. This in turn leads to an important differentiation between the point of departure at, on the one hand, the conceptual level, and the point of departure at the econometric level, on the other hand. In the former it is the *concept of exchangeable value*, while in the latter it is the *actual prices given as data*.

The assumptions (a) – (d), once unified with (w_1) and (w_2), provide the following econometric yield.

Let us take the price P of the same physical amount of a certain commodity at two distant moments in time, t_1 and t_2, $P(t_1)$ and $P(t_2)$. Based on the knowledge of these two prices, one can, first, compute their difference $P(t_1) - P(t_2)$ and, then, using the data on the prices of corn at t_1 and t_2, compute the respective

---

[152] The role of labor embodied in determining value in the context of Smith's theory is notoriously controversial. On the one hand, Smith states that "in a civilized country there are but few commodities of which the exchangeable value arises from labour only, rent and profit contributing largely to that of the far greater part of them" (vi. 24). On the other hand, as O'Donnell points out, Smith seemed unhappy with the idea of profits and rents as "sources of value" (O'Donnell 1990, 88). For the purposes of our analysis, we rest content – until the final part of our chapter – with O'Donnell's conclusion that Smith "considered prices to be determined by methods of production and took *as given* the prevailing rates of wages, profits and rents" (O'Donnell 1990, 91). Under this assumption, changes in the amount of labor embodied result in proportional changes in value: "[A] constant quantity of labour expended in production creates a constant quantity of value." (O'Donnell 1990, 65).

difference in the amounts of corn. This difference represents the difference in labor commanded by the commodity at t_1 and t_2. As it is proportional to the change in the labor embodied between t_1 and t_2, it can be used to compute this change, which ultimately has its roots in changes in the methods of production. This interlocked sequence of computations, as well as the intervening assumptions on which these computations are based, is expressed in the following diagram:

Figure 10.1: The interlocked sequence of econometric computations as based on Smith's theory and assumptions w_1 and w_2

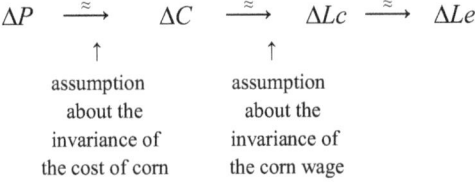

Here, ΔP stands for the difference in prices at t_1 and t_2; ΔC for the difference in the amount of corn; ΔLc for the difference in the amount of labor commanded; ΔLe for the difference in the amount of labor embodied; the symbol "$\approx$" in "$\xrightarrow{\approx}$" stands for the proportionality of quantities represented by the symbols on both sides of this sign, while the arrow indicates the direction in which the computation is performed; the vertical arrows indicate the assumptions under which the first two computations can be performed.

Let us now turn to two charges raised against Smith by his critics, old and new. *First*, it is sometimes claimed that Smith confuses the *cause* and the *measure* of exchange value. This charge does not hold. In Chapters V and VI of Book I, Smith explicitly holds to the view that quantities of labor embodied (co-)determine[153] the ratio in which they are mutually exchanged. Moreover, once the qualitative-*cum*-causal assumption (d) is taken into account, it can be readily seen that there is also a qualitative dimension to Smith's theory, namely that *the direction of causation goes from labor embodied to exchangeable value*. An economic characteristic in its qualitative aspect is the basis for the formation of another economic characteristic as its effect, this effect being produced in the course of the mutual interaction of commodities in the process of exchange. The econometric computation outlined above is based on a conceptual

---

153   See footnote 152 above.

understanding of the direction of causation, but goes in the *opposite* direction, as shown in the diagram. *The knowledge of the effect* (as expressed in the data about its quantitative characteristics) *enables* (under the given assumptions and via the conceptual chain described above) *the computation of the quantity of its cause.*

Second, it might seem that Smith confuses the concept of labor commanded with that of labor embodied or vacillates between them. But as shown above, Smith views them as distinct and only claims that they are quantitatively proportional under certain assumptions, so that based on the knowledge of the difference in the quantity of the former one can infer the difference in the quantity of the latter.

J. S. Mill expresses these two points in his defense of Smith in the *Principles* as follows ([1848] 1965, Vol. II, 100–101):

> The idea of a Measure of Value must not be confounded with the idea of the regulator, or determining principle, of value. When it is said by Ricardo and others, that the value of a thing is regulated by quantity of labour, they do not mean the quantity of labour for which the thing will exchange, but the quantity required for producing it. This, they mean to affirm, determines its value; causes it to be of the value it is, and of no other. But when Adam Smith and Malthus say that labour is a measure of value, they do not mean the labour by which the thing was or can be made, but the quantity of labour which it will exchange for, or purchase; in other words the value of the thing, estimated in labour. And they do not mean that this *regulates* the general exchange value of the thing, or has any effect in determining what that value shall be, but only ascertains what it is, and whether and how much it varies from time to time and from place to place.

Mill also evaluates the possible confusion in these matters using an example from physics, namely: "To confound these two ideas, would be much the same thing as to overlook the distinction between the thermometer and fire" (Mill [1848] 1965, Vol. II, 101). But what shall one make of this evaluation? And wherein, exactly, lies the analogy between the relation of the fire to the thermometer and the relation of (Smith's and Malthus') measure of value to (Ricardo's) quantity of labor as the regulator of value? The answer to this question can be formulated by means of a reconstruction of *categories* explicitly or implicitly at work, at the meta-conceptual level, in Book I of *WoN*.

Such a reconstruction can be justified in the following way. First of all, in Book I, Smith himself employs certain terms, namely "measure" and "standard". These can be viewed as elements of a general framework within which Smith addresses issues of economic quantification and econometric measurement. Also worth noting is Smith's use of the category of "cause". It appears not only in the title of his 1776 oeuvre, but also in the title of Book I: "Of the causes of improvement in the productive powers of labour". Book I can be viewed as an investigation into the causal mechanism at work in the economic structure of modern commercial societies. An analysis which identifies the set of categories involved in Book I could thus provide a general and at the same time an in-depth understanding of how this investigation proceeds. Second, such a set of categories can serve as an instrument in comparing Smith's Book I with other major works in political economy, provided, of course, that these are also subjected to a meta-conceptual analysis focusing on the categories involved. The following section is the first attempt at this.

10.3 Measure, Standard and Cause: Meta-conceptual Reflections

Let us now turn to the concepts from Smith's Book I that are present in the quotation from Mill's *Principles*. First of all, what is the epistemic status of the concept of exchangeable value? It is first introduced in Chapter IV in the context of the plan of investigation, after Smith's analysis of the division of labor and exchange. With respect to the concept of labor embodied, it represents the starting point of a sequence of concepts and their derivations, the aim of which is to identify and quantify the former. Also, Smith's initial understanding of the concept of exchangeable value ("the power of purchasing other goods which the possession of that object conveys") draws on the everyday knowledge of agents involved in the exchange of commodities, that is, on how the process of exchange *appears* to them. For these reasons, we assign to this concept the category of *appearance*.

Earlier, we characterized the relation of labor embodied and exchange value as causal. The character of labor embodied as a *cause*, with respect to exchangeable value, can now be further specified by assigning to the concept of labor embodied the category *ground* underlying the *appearance*.

Let us now turn to Smith's understanding of measure. As shown above, Smith proceeds at the conceptual level from the ratio of the quantities of exchangeable values of a commodity at times t_1 and t_2 to the ratio of the quantities of labor embodied (i.e., to changes in the methods of production between t_1 and t_2). Once we view the relation between exchangeable value and labor embodied as that of an effect to its cause, Smith's approach can be understood as based on the view that *the knowledge of the quantity of effect/appearance should enable one to derive the knowledge about the quantity of the cause/ground underlying the appearance*. With respect to this approach, the concept of exchangeable value can be characterized by the category *external measure*, which expresses, at the meta-conceptual level, Whewell's second axiom of reasoning about causes, i.e., *"Causes are measured by their effects"* (1834, 151).

With this in mind, it is now possible to explain why Mill invoked the example of a fire and a thermometer. Fire – or, more precisely, heat – is the *cause* which produces as its *effect* a change in the thermometer reading, while, conversely, the values of this reading can be used to compute the amount of heat actually at work in this change.

As shown above, Smith presupposes that the ratios in which different quantities of certain kinds of commodities are exchanged have their basis and origin in the quantities of labor embodied in them. We hypothesize that this is the reason why Smith calls labor the "*real* measure" of value and the "*real*" or "*ultimate*" standard of value. To the quantitative characteristic of labor embodied (the amount of "toil and trouble", i.e., a quantity of expended labor time) we can assign the category *immanent measure*: the quantity of labor embodied as the cause/ground *determines the quantitative characteristics* of the phenomenon it produces.

As we have seen, Chapter V unifies, on the one hand, data-driven considerations of the variations of concrete quantitative determinations of economic characteristics of commodities like precious metals and corn with, on the other hand, more abstract reflections of concepts like labor commanded, exchangeable value, etc., as well as their relations. The latter, more general and "theoretical" reflections are not concerned with economic measurement or empirical data *per se*; instead, they yield new concepts or new relations between pre-existing concepts. However, they are still necessary to Smith's econometric concerns: the construction of new relations between the concepts of the corn price, labor commanded and labor embodied is precisely what enables Smith to establish a link between variations in exchange ratios and variations in methods of production. Together, the two dimensions of Chapter V – the econometric and the theoretical – are put to use by Smith to econometric purposes in later chapters of *WoN*.

Once this co-existence of an econometric and a theoretical dimension of Chapter V is explicitly acknowledged, it is easy to recognize the following three *illusions* which may arise in the analysis of this chapter, or of Book I in general. *First*, the illusion that Smith's treatment of concepts like value, labor, etc., *necessarily* implies that certain commodities (e.g. corn) should have time-invariable quantitative economic characteristics. *Second*, the illusion that once one subscribes, at the theoretical level, to the view that the quantity of value of a commodity is determined by the quantity of labor embodied or "contained" in the commodity, then the actual amounts of these quantities should be computable at the level of econometrics. And *third*, the illusion that once such attempts at an econometric computation of these amounts fail, then the theory from which this requirement of computability was (seemingly) derived is refuted.

The distinction between the theoretical and econometric dimension of Chapter V can also be addressed from the point of view of categories. At the meta-conceptual level, we may distinguish the *quantity* (the initial, indeterminate quantitative characteristic) and the *quantum* (a determinate quantitative characteristic) of an object under investigation. The category of quantity is at work in the derivation of quantitative determinations of objects of investigation from their other quantitative determinations. For example, Smith, as shown above, derives the ratio of *quantities* of labor embodied in commodities from the ratio of the *quantities* of exchangeable values of commodities. It is, however, the practical task of econometrics to attempt to compute the actual *quantum* (amount) of, say, labor embodied in a particular commodity of a certain type, based on the knowledge about the *quantum* of its market price. And in order to perform such a computation, one is in need of a *standard* of prices which is either constant over time or can be adjusted to provide reliable data across distant moments in time. Thus, what is in everyday terminology called *measurement* is in fact based on the unification of two related, but nonetheless different aspects of cognition, characterized by the different categories of *quantity* and *quantum*.

These two aspects can also be distinguished by means of the category-pairs *measurans-measurandum* and *computans-computandum*. At the level of general economic theory, it is possible to move from the knowledge about a quantity of a magnitude to the knowledge about the quantity of another magnitude, where the former has the status of a *measurans* and the latter the status of a *measurandum*. However, only at the level of actual measurement – i.e., in the case of political economy, econometrics – can the knowledge about the *quantum* (amount) of the former magnitude serve as the *computans* in the derivation of the *quantum* of the other magnitude, which would then have the status of a *computandum*. The two levels, econometric and theoretical, co-exist in Smith's *WoN* and are closely related. Generally, however, the presence of the pair *measurans-measurandum* in a particular conceptual system of political economy does not *necessarily* imply that the knowledge of the actual price of a commodity should enable the computation of the quantum, that is, the actual amount of labor embodied in that commodity.[154]

---

154 Marx, for example, held that on the basis of his theory of value, "The labour applied to the individual commodity can no longer be computed at all" (1988a, 33).

Our sketch of a meta-conceptual reconstruction of Smith's endeavor enables certain clarifications of both *WoN* and its more recent interpretations. Following Smith, Henry (2000, 2) states that "for Smith labor was both the cause and the measure of value". But while labor embodied is, for Smith, the *cause* of exchangeable value, it can, strictly speaking, be neither the *measure* of exchangeable value, nor the measure of exchangeable *value*. Only *quantities* of labor embodied can be viewed as determining the *quantities* of exchangeable value. Thus, only quantified "toil and trouble", once understood as, i.e., the labor time (the duration of labor), can be said to be the (immanent) measure of value.

Henry (2000, 4) also states that for Smith, "labor is the standard by which the value is determined". Smith himself refers to labor as "the ultimate and real standard" of value. However, a closer look at the relevant passages reveals that for Smith, it is not as *standard* that labor *determines* value. Labor as standard allows for the *comparison* and *estimation* (v. 7) of values, as do other standards (like precious metals), albeit less reliably. It is as cause that labor determines value. Similarly, in the conceptual sequence reconstructed above, the ratio of quanta of exchangeable values of commodities serves, under certain assumptions, as a *standard* in the computation of the ratio of the quanta of labor embodied in these commodities.

10.4 Conclusion: Concepts, Categories and Smith's Natural Price

So far, we have focused on the sequence of economic concepts in Chapter V and its econometric implications as represented in Diagram 1. Let us now approach this chapter in the wider context of Book I.

In Chapter IV, Smith analyzes the employment of coined money as the medium of exchange and addresses its development in various stages. From there he proceeds, via Chapter V, to Chapter VI. In this latter chapter, Smith draws on the assumption – reflected upon already in Chapter V – that the quantities of labor embodied in a commodity and of labor commanded by a commodity differ. This quantitative difference is now understood as originating in the magnitudes of profit and rent which enter, in addition to wage, into the structure of the price of a commodity. Finally, in Chapter VII, Smith transforms the concepts of wage, profit and rent into those of their averages. This, in turn, leads to the transformation of the concept of price into that of natural price, and to the transformation of actual price into that of market price.

This overall sequence of economic concepts, stretching from Chapter IV to Chapter VII, can be represented as follows:

Figure 10.2: The sequence of Smith's economic concepts from Chapter IV to Chapter VII of Book I of WoN

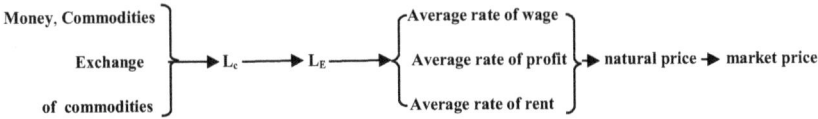

This sequence displays an important feature: the concepts at the end of the sequence enable Smith to treat certain economic problems which were not and could not have been treated at its beginning – for example, the distribution among the classes of laborers, manufacturers and landowners. However, for Smith's attempt at construction to be entirely successful, the sequence of concepts (more precisely, of the definitions and re-definitions of these concepts) must be internally coherent. In other words, the conceptual components of the sequence must fit seamlessly. We argue that this is not the case.

The conceptual incoherence we have in mind pertains to the role of the concept of labor embodied in the sequence. It unifies and mediates between two phases of a conceptual movement: *from* concepts like money and exchange, *to* that of labor embodied, and *from* the latter, via the concepts of the average rates of wage, profit and rent, *to* that of natural price and, finally, of market price. However, in Smith's approach, profits and rents (as well as their averages) are not caused by labor embodied. He states this explicitly with respect to profit: "Neither is the quantity of labour commonly employed in acquiring or producing any commodity, the only circumstance which can regulate the quantity which it ought commonly to purchase, command, or exchange for" (vi. 7).

Smith's view of profit and rent as not based on labor embodied subjects his econometric endeavor to the burden of purely contingent conditions which are, however, required to compute the difference in labor embodied from the difference in price. We can illustrate this on the following example of measurement, in which the economic magnitudes of prices are given as input data for the computation of labor commanded and the change in labor embodied.[155]

---

155 In square brackets, we specify the units of measurement in which the quanta of magnitudes are expressed. To the best of our knowledge, Smith never assigned time units to quanta of labor embodied. Under "pure number" we understand a number devoid of any unit of measurement; under "denominated number" we understand a number unified with a unit of measurement. The price relations expressed in the table are purely fictitious.

Table 10.1: Example of the data and constants, both known and unknown, involved in the computation of the change in time of labor embodied

Time	Economic magnitudes & their quanta / Data / Price of a commodity at time t_1 [£]	Constant / Cost of corn per lb. [£]	Constant / Corn wage per working day [lbs.]	Constant / Ratio of surplus share to price of commodity at t_1 [unknown, pure number]	Computed / Labor command of commodity at t_1 [working days]	Computed / Labor embodied in commodity at time t_1 [working days; actual denominated number unknown]
t_1	200	x	y	a	4	z
t_2	100	x	y	a	2	$\dfrac{z}{2}$

The variables x and y in the third and fourth columns, respectively, acquire values which do not change and where these values are accessible as data. From the given prices, the cost of corn and the corn wage, one can compute the labor commanded by a commodity at the given time: (200 £ ÷ 10 £ per lb.) ÷ 5 lbs. per working day = 4 working days commanded at time t_1. Comparing the results for t_1 and t_2, one can see that the labor command of the commodity has halved. Supposing (w_1) and (w_2), this change in labor commanded will be proportional to the change in labor embodied due to changes in methods of production. Hence, even though the actual amount of labor embodied at t_1 is unknown (expressed by the variable z in our table) because it cannot be computed, based on the former computation, one can compute at least that the labor "contained" in the commodity has been halved between t_1 and t_2 as well.[156]

Let us now turn to the fifth column. As a result of the nature of profit and rent in Smith's theory, the computation above is only feasible under the condition that the ratio of overall surplus (that is, profit plus rent) to the price of the commodity does

---

156 One might object that this is already obvious from the prices themselves. In our example, the ratio of the price at time t_1 to the price at t_2 is indeed equal to the respective ratios of labor commanded and labor embodied. However, this is only because we presuppose, for the sake of simplicity, that the prices given as data were adjusted for changes in the value of money.

not change between t_1 and t_2. The actual quantity of this ratio is unknown, but presupposed as constant; we denote it as a. Had *this ratio been subjected to a change*, then, the changes in labor command would no longer be proportional to changes in labor embodied. Let us suppose that the labor embodied in the commodity at t_2 would be halved as in our table, but the surplus share would increase to $2a$ (leaving the other constants unchanged). The resulting change in labor commanded, as computed from the price, would not reflect the actual change in labor embodied.

Smith's interlocked sequence of computations expressed in Figure 10.1 is thus shown to rest on yet another assumption (s_1) of a constant surplus share in the price of the commodities involved:

Figure 10.3: Smith's sequence of computations revised to include the assumption s_1

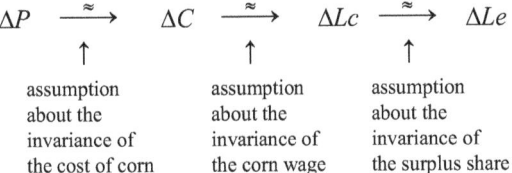

$$\Delta P \xrightarrow{\approx} \Delta C \xrightarrow{\approx} \Delta Lc \xrightarrow{\approx} \Delta Le$$

↑	↑	↑
assumption about the invariance of the cost of corn	assumption about the invariance of the corn wage	assumption about the invariance of the surplus share

Finally, let us turn to the conceptual sequence represented in Figure 10.2 in order to note its peculiar character. Not only does it enable the treatment of certain issues which was not possible at the beginning; it also enables a more precise understanding of the concepts present at its beginning. For example, while Smith, at the beginning, merely presupposes that commodities are exchanged, he can, at the end, further clarify the process of exchange at the end of the sequence, by means of the concepts of natural price and market price. In other words: Smith's conceptual sequence has a *cyclical* character in that the latter parts "feedback" to the former parts.

How can this cyclical character of a sequence of concepts be approached from the point of view of categories? C. G. Hempel, in his *post*-logico-empiricist phase, characterized it as follows (Hempel 1970, 142):

> Theories are normally constructed only when prior research in a given field has yielded a body of knowledge that includes empirical generalizations or putative laws concerning phenomena under study. A theory then aims at providing a deeper understanding by construing those phenomena as manifestations of certain underlying processes governed by laws which account for uniformities previously studied.

Above, we assigned to the concept of exchangeable value at the beginning of the cycle the meta-conceptual status of *appearance*. Drawing on Hempel's terminology, we may assign to the concepts at the end of Smith's sequence (as represented

261

in Figure 10.2) the meta-conceptual status of *manifestations*. The whole cycle from phenomena as appearances to phenomena as manifestations, i.e., the sequence of concepts in Chapters IV through VII of Book I can then be characterized, in the language of categories, as the *retreat from appearances and going into their underlying mechanism (ground/cause), then, coming out from the latter and, finally, going to the manifestations.*

Chapter 11: Open Problems: Categories, Types of Thought Objects and the Historical Method

In the previous chapters I dealt in various ways with the scientific, metascientific, and methodological levels of the issue of scientific explanation. Thus, in Chapter 1 I proposed a variety of possible explanation procedures, some of which involved heuristic moments. Even if these procedures differed from one another, still they shared a common feature, namely, that of *mediation*. For example, Figure 1.3 represented the situation when the derivation of an antecedently given scientific law from an *explanans* law requires the discovery of modification conditions which can serve as mediating links between the *explanans*-law and the *explanandum*-law.

Similarly, in Chapter 2, the whole sequence represented in Figure 2.2 can be viewed as a sequence of mediating links involved in the development of cognition from data to the laws of manifestations and individual manifestations. In Chapter 7 I dealt with Tomasello's three-phase evolutionary hypothesis, where Phase 2 should serve as a mediating link between Phase 1 and Phase 3. Finally, Chapter 8 reconstructed Marx's derivation of the various forms of profit as being accomplished on the basis of the law of value plus the introduction of mediating categories like surplus value, constant capital, and variable capital.

Accordingly, I view these explanation procedures as a variety of thought operations, part of which is also the introduction of the respective mediating links. Let me now try to find out what types of thought objects are involved in these various explanation procedures.

11.1 A Typology of Thought Objects for Types of Scientific Explanation

In Chapters 1, 2 and 3 I dealt with thought constructs like the mathematical pendulum and the idealized (i.e., nonrelativistic) hydrogen atom. Idealizations were also involved also (as shown in Chapter 6 and Chapter 7) in practical experiments with great apes and human infants. For example, as shown in part A) of 7.2.2, the tests trials dealt with the ability of great apes and children to approach in an innovative way a problem situation under conditions that were in thought initially presupposed as not operative, and then experiments were performed where these idealizations were implemented (i.e., a certain type of thought closure and then experimental closure was at work).¹⁵⁷ In a next step, these idealizations/closures where given up first in mind (cancelled, abolished) and then in experimental trials. By taking into account these idealizations/closures, how they were initially introduced and then given up, one can call the entities, initially given as thought-entities and then experimentally constructed as *idealized entities*.

A different type of thought object is involved in the *structural-genetical method* of research/explanation method dealt with in part B) of 7.2.2. Here the focus is on the formation of the structure of an object under investigation, that is, this structure – contrary to the structure of the idealized object – displays a certain developmental *diachronicity*. At the same time all the particular objects whose formation led, due to the impact of secondary factors (modification conditions), to a different structure are viewed only as subtypes of the former which is understood as an exemplifying a *classical type of thought object*.

This classical type of thought object also seems to appear in Marx's *Capital* where the formation process of the various forms of profit is explained on the basis of the law of value. In the manuscript of Book III, when dealing with the obstacles that hinder the equalization of wages and thus of the rate of profit between different branches in one country, he refers to the type of thought object being involved here as follows (1992, 215):

> Important as the study of such frictions is for any special research into labor-wage, they are to be removed (to be ignored) as accidental and nonessential. In such a general investigation it is always presupposed that the real relations correspond to their concept or, what is the same, [that] real relations are represented only insofar as they express (represent) their own general type.

And when emphasizing in this manuscript that he does not deal with the impact of the world market, the movement of the market prices, the business cycles, etc., he justifies this as follows: "This is because the real movement of competition lies outside of our plan, and we have to represent only the inner organization of the capitalist mode of production, (represent it), so to say, in its ideal average" (1992, 853).

But once the focus of research is on the differences in the structures of the various subtypes, then each of them turns into an *original type* of a thought object. So, for example, when Tomasello compares typically developing human infants with autistic ones, then one can regard the separate theoretical treatment of each of

---

157   In this employment of idealizations as closures I draw on (Pawson 1989). The term "closure" goes back to (Bhaskar 1978).

them as pertaining to an *original thought-type in its own right*. My introduction of this type of thought object draws on Marx's reflections on the thought derivation of descriptions of various historical types of properties. He declares about the naturally arisen (*naturwüchsig*) communal propert as follows (1980a, 113):

> It is the original form (*Urform*), which we can be found among Romans, Teutons and Celts, but of which a whole sample-card with diverse specimens can still be found, even if partially in the form of ruins, in India. A detailed study of the Asiatic, especially Indian communal property, would prove how from the different forms of the naturally arisen communal property arise the different forms of its break up. Thus, for example, the various original types of Roman and Teutonic private property can be derived from the various forms of Indian communal property.

My differentiation among the various types of thought object, when combined with my reflections in Chapter 7 on the methods of research/explanation as given in the works of Tomasello *et alia* yields the following table.

Table 11.1: Overview of methods applied or not applicable in Tomasello's developmental and comparative psychology

Method	Thought object(s) in the method	Relation between thought objects in *explanans* and in *explanandum*	Tomasello's application of the method in research/ explanation of
Structural	Idealized object	No time diachronicity in the structure of idealized object	Structure of social-cognitive skills in great apes and in human infants/children
Structural-genetic	Classical type	Diachronicity inside the structure of the classical type (its formation)	Ontogeny of cooperative communication in typically developing human infants/children
Structural-historical	Original types	Hypothetical formation of structure of an original type from the structure of another one taking place in hypothetical past	Hypothetical phylogeny of cooperative communication *Apes→ Homo→ Earlier sapiens→ Later sapiens*

Method	Thought object(s) in the method	Relation between thought objects in *explanans* and in *explanandum*	Tomasello's application of the method in research/ explanation of
Genetic-historical	Original types	Real formation of structure of an original type from another one taking place in real past	**Absent because not applicable** Real phylogeny of cooperative communication in Last Common Ancestor of Great Apes and Homos→Homo heidelbergensis→Homo sapiens sapiens

The assignment of the names "*Homo*," "*Early Sapiens*," and "*Late Sapiens*" from Table 7.6 in Chapter 7 to names of species in an evolutionary tree would turn Tomasello's quasi-evolutionary story in this table into a more complete or even a fully blown one. But this turn depends crucially not only on the completeness and exactness of the reconstruction of such an evolutionary tree, but also on that assignment and, as I shall now show, on the fact that this assignment cannot be accomplished because of the specific subject-matter at which Tomasello's Table 7.6 aims.

The subject-matter of Tomasello's research are the cognitive/linguistic capabilities of great apes and of human infants/children. As shown above, he starts from a conceptual reconstruction of the ontogeny of typically developing human infants/children, combined with research into autistic children, children with developmental delays, and great apes, and then provides a highly hypothetical scheme of a phylogeny-spanning genera like apes and *homo* and species like *Homo sapiens*. And, as shown above, he does not conceptually specify in detail the extinct ape-species and *homo*-species, in which the respective cognitive/linguistic skills should have been given. The absence of this specification has its roots in the fact that the conceptual reconstruction of the sequence of hominin species is based primarily on the discovery of fossilized bones and an analysis of their morphology, genetic characteristics, etc. But such an analysis does not enable one to derive a conceptual description of the cognitive/linguistic skills of the respective hominin species.[158]

---

158 On problems in the explanation of the relation of cognition to genes see (Fisher 2006).

This inability can be expressed even in even more exact terms. The presence of certain characteristics discovered in the fossils, like the shape and interconnection of bones in the hand,[159] or the presence of the hyoid bone, do not determine unambiguously that the individuals from the extinct species produced certain types of stone tools or communicated via speech.[160] Social skills and behaviors "do not fossilize" (Shultz and Christopher and Atkinson 2011, 219). Neither do those characteristics determine the semantics of their (possible) language communication, say, that it was already propositionally differentiated. The absence of an unambiguous determination of the (possible) presence of cognitive/linguistic skills is exactly that which was expressed by Tomasello as their *enabling but not determination by biological characteristics*.[161]

11.2 The limits of the applicability of the category cluster *appearance – essence (ground) – manifestation*

In chapters 1 through 10, I dealt with the employment of the following triad: *appearance – essence (ground) – manifestation* in *scientific thinking about the things of nature and/or society*. This tripartite cluster, however, does not correspond to one possible type thinking in theoretical science, namely, when the latter tries think the *origin of a ground with a specific mechanism* in the sense of its origin in other essences (grounds) with different mechanisms. The categories in those chapters correspond only to such type of thinking which understands these things as based on an already *existing* and simultaneously *enduring* mechanism of the ground.

The type of deficit on the side of categories I have in mind here can be explicated with respect to the article (Glennan 2010). Glennan differentiates between the category *historical explanation* and the category *ahistorical explanation* as follows (2010, 252–253):

---

159 For an attempt to found out – by subjecting human muscles employed in stone tool production to electromyography – if individuals from certain extinct species were able to produce Oldowan style instruments, see (Marzke *et al.* 1998).
160 On this see (Lieberman 2007).
161 On this see (Tomasello 1999a, 11, 19, 53, 90).

[Historical explanation]	[Ahistorical explanation]
I will take it to be the defining characteristic of an historical explanation that it explains the occurrence of some *particular* event or state of affairs by describing how it came to be.	Ahistorical explanations abstract from particular events at particular places and times in order to explain recurrent patterns of phenomena … one does not explain a particular event, but some sort of repeatable phenomenon … the same explanation works for potentially countless instances of that kind of phenomenon.

And he also states the following (2010, 258).

> Systems mechanisms have a historical dimension in the sense that generalizations describing the behavior of … mechanisms are true only in virtue of the particular organization and interaction of parts, and of the historical processes that brought them about … But while the fact that there always be a history that explains why a certain mechanism exists and behaves as it does, one need not know that history to explain how the mechanism works … Thus, mechanistic explanations themselves appear to be ahistorical. They describe the regular operation of reliable mechanisms.

The following issue is worth noting here. When Glennan uses the category *historical explanation*, he targets only the description of the occurrence of a particular event or state of affair as the thought object involved in it; that is, he bypasses the issue of how to deal with *thinking targeting the history of the mechanism of the very ground*. Under the category *history of the mechanism of the very ground* I understand the *formation* of a ground's mechanism with its constituent entities and activities out of another, declining mechanism with its own specific constituent entities and activities, and the decline and disappearance of a mechanism due to its transformation into another mechanism. Stated in the terminology of categories this means that under *historical explanation* one can understand not only a thought movement from the *explanans*-description of the working of a mechanism to the *explanandum*-description of a particular event produced by this mechanism, but also as a thought movement from an *explanans* to an *explanandum* where *both* stand for a description of a ground's mechanism, mutually, of course, differing in the described entities, activities, etc.

There are in fact two issues involved in what Glennan labels by the category *historical explanation*. One, when the aim of the thought derivation is to explain how a system with its specific mechanism is subjected to an action from the outside, say, by a system with another mechanism, and where in the course of this action in the former system certain events are produced, or, even its mechanism collapses under the impact of this action, and so the system subjected to the action disappears. To this latter case correspond, for example, the attempts to explain the event of extinction of the dinosaurs by the impact of a huge asteroid.

But the goal of historical explanation can be also to understand how a mechanism is formed from a temporally earlier mechanism and also how and why did it declined and eventually disappear by transforming itself into another, temporally later mechanism, and where the cause of the formation and decline of the mechanism is explained by thinking about the workings of the very mechanisms themselves. Thus, despite the reconstruction and employment of a cluster of categories in the chapters of this book, I did not provide a cluster of categories which would enable one to think the things of nature and/or society *not only as given and as enduring, as well as historically formed and historically declining.*

This deficit also can be approached by turning to the category triple of *appearance – essence (ground) – manifestation* and to the method of the *ascent from the abstract to the concrete* dealt with from the point of view of Marx's realistic epistemology in Chapter 8. Both this triple and method can serve the purpose of a thought-reproduction as a thought concrete of a concrete existing outside thinking. Thus, they, first, presuppose the existence of a real concrete and subsequently enable its thought reproduction; but they do not make it possible to investigate into and explain that historically prior concretes' from which it developed. Such, in any case, is my interpretation of the following passage from Marx's *On the method of political economy*. Initially he states the following (1976, 40):

> The bourgeois society is the most developed and most diversified historical organization of production. The categories which express its relations, the understanding of its inner division (*Gliederung*), provide therefore at the same time also an insight into the inner division and relations of production of all those defunct societal forms out of whose ruins and elements it has built itself, from which partially still not overcame remnants are dragged along within it, whose mere indications have developed into fully fledged importance, etc. The human anatomy is the key to the anatomy of the ape. The indication of a higher (*ein Höhres*) in less developed animal species can only be understood, if the higher is already known. The bourgeois economy provides thus the key to the ancient, etc.

But, at the same time, he *cautions against mixing up the understanding of the inner structure of a real concrete with the understanding of the inner structure of historically antecedent real concretes from which it developed*: "Although ... it is true that the categories of the bourgeois economy possess a truth for all other societal forms, this is to be taken only *cum grano salis*. They can contain them developed, stunted, caricatured, etc., always in essential difference" (1976, 40). This caution thus requires that Marx's claim "The human anatomy is the key to the anatomy of the ape" be complemented with the amendment, "but it cannot replace the latter's anatomy."[162]

From this I draw the conclusion that we *are need, in order to deal theoretically with the issue of the origin of a real concrete from other real concretes, of a new set of categories and a new set of methods of research and explanation*. The justification for such a need can also be given when one takes into account Marx's typology (given in a letter to Engels of April 2, 1858) of possible objections against the claim that value has no other "substance" ("*Stoff*") than labor. This typology is as follows (2003c, 123):

> All objections to this definition of value are taken either from underdeveloped relations of production, o[r] are based on the confusion, where the more concrete economic determinations from which value is abstracted (a[nd] which therefore may also be seen, on the other hand, as a further development of the same) are upheld against it in this its abstract, underdeveloped form.

So, a critique of these two types of objections has to draw on two different types of reasoning. As to the second type of objection, its critique has to draw on reasoning based on the method of the ascent from the abstract to the concrete unified with the trio of categories *appearance – essence (ground) – manifestation*. This critique aims to show that objections of this type fail because they try to refute claims about the essence (ground) of some kinds of entities by means of claims about their apparent phenomenal characteristics. As to the first type of possible objections, its critique has to draw on the method enabling the thought movement between different thought concretes differing in their respective grounds (essences) and where this method would be unified with a cluster of categories different from that of *appearance – essence (ground) – manifestation*. This critique should show that objections of this type fail because they try to refute claims about the essence (ground) of some kinds of entities by means of claims pertaining to entities of another kind with a different essence (ground).

It remains the task of a future project to provide a method for enabling the thought movement between different thought concretes which differ in their respective grounds (essences) and a cluster of categories corresponding to this method. Here I can only indicate the conceptual resources from which this task would likely draw.

The first conceptual resource should be the more recent philosophy of science dealing with the issue of explanation in the historical natural and social

---

162   Here I draw on (Černík 1978).

sciences.[163] The second conceptual resource should be those philosophical doctrines which provide at least some philosophical, though yet to be to be identified, categories of historical cognition. The third conceptual resource should a set of empirical historical (natural and/or social) sciences which would function as a "counterbalance" to the first two conceptual resources. As prime candidate I would consider modern paleontology (especially modern paleogenetics in which human genome serves as the key to the genomes of the previous extinct *Homo* species without, of course, functioning as a substitute for the reconstruction of their respective genomes).

---

163 On explanation in historical natural sciences see (Cleland 2002), (Cleland 2011), (Jaffares 2010), (Turner 2005) and (Turner 2007).

References

Albiach-Serrano, A. and Call, J. and Barth, J. 2010. Great Apes Track Hidden Objects after Changes in the Objects' Position and in Subject's Orientation. *American Journal of Primatology* 72: 349–59.

Ångström, A.-J. 1868. *Recherches sur le spectre solaire*. Upsal: W. Schultz.

Balmer, J. J. 1884. Wellenlänge der Spectrallinien des Wasserstoffs. [Nine page manuscript deposited in the archives of the library of the University Basel, Switzerland].

Balmer, J. J. 1885a. Notiz über die Spectrallinien des Wasserstoffs. *Verhandlungen der Naturforschenden Gemeinschaft in Basel* 7: 548–55.

Balmer, J. J. 1885b. Zweite Notiz über die Spectrallinien des Wasserstoffs. *Verhandlungen der Naturforschenden Gemeinschaft in Basel* 7: 750–52.

Baly, E. C. C. 1912. *Spectroscopy*. London: Longman, Green, and Co.

Banet, L. 1966. Evolution of Balmer Series. *American Journal of Physics* 34: 496–503.

Banet, L. 1970. Balmer's Manuscripts and the Construction of His Series. *American Journal of Physics* 38: 821–28.

Batterman, R. W. 2009. Idealization and Modeling. *Synthese* 169: 427–46.

Batterman, R. W. and Rice, C. C. 2014. Minimal Model Explanation. *Philosophy of Science* 81: 349–76.

Bechtel, W. 2011. Mechanism and Biological Explanation. *Philosophy of Science* 78: 533–57.

Beller, M. 1999. *Quantum Dialogue*. Chicago: The University of Chicago Press.

Benson, J. D. et al. 2004. Mind and Brain in Apes. *Language Sciences* 26: 643–60.

Beran, M. J. 2004. Chimpanzees (*Pan troglodytes*) respond to Nonvisible Sets after One-by-One Addition and Removal of Items. *Journal of Comparative Psychology* 118: 25–36.

Beran, M. J. et al. 1998. Symbol Comprehension and Learning. *Evolution of Communication* 2: 171–88.

Berka, Karl. 1983. *Measurement*. Dordrecht: Reidel.

Bevan, P. W. 1910. Dispersion of Light by Potassium Vapour. *Proceedings of the Royal Society of London A* 84: 209–25.

Bhaskar, R. (1978): *A Realist Theory of Science*. Atlantic Highlands (NJ): Humanities Press.

Biro, D. 2011. Clues to culture? In: Matsuzawa and Humle and Suguyami eds. 2011: 201–10.

Biro, D. and Carvalho, S. and Matsuzawa, T. 2010. Tools, traditions and technologies. In: *Chimpanzee Minds*, eds. E. V. Lonsdorf and S. S. Ross and T. Matsuzawa, 141–55. Chicago: Chicago University Press.

Birtwistle, G. 1926. *The Quantum Theory of Atom*. Cambridge: Cambridge University Press.

Birtwistle, G. 1928. *The New Quantum Mechanics*. Cambridge: Cambridge University Press.

Black, J. 1807. *Lectures on the Elements of Chemistry*, Vol. I. Philadelphia: Mathew Carey.

Blumberg, A. and Feigl, H. 1931. Logical Positivism. *Journal of Philosophy* 28: 281–96.

Boerhaave, E. 1743. *A New Method of Chemistry*. London: Longman.

Boesch, C. 2002. Cooperative Hunting Roles Among Taï Chimpanzees. *Human Nature* 13: 27–46.

Boesch, C. 2013. Ecology and cognition of tool use in chimpanzee. In: *Tool Use in Animals*, eds. C. M. Sanz and J. Call and C. Boesch, 21–47. Cambridge: Cambridge University Press.

Boesch, C. and Boesch, H. and Vigilant, L. 2006. Copperative hunting in the chimpanzees. In: *Cooperation in Primates*, eds. P. M. Kappeler and C. P. Van Shaik, 139–50. Berlin: Springer.

Boesch, C. and Head, J. and Robbins, M. M. 2009. Complex Tool Sets for Honey Extraction among Chimpanzees in Luongo National Park, Gabon. *Journal of Human Evolution* 56: 560–89.

Bogen, J. 2011. 'Saving the Phenomena' and Saving the Phenomena. *Synthese* 182: 7–22.

Bogen, J. and Woodward, J. 1988. Saving the Phenomena. *Philosophical Review* 97: 303–52.

Bogen, J. and Woodward, J. 1992. Observations, Theories, and the Evolution of Human Spirit. *Philosophy of Science* 59: 590–611.

Bohr, N. 1911. [Lecture on the Electron Theory of Metals (13 November 1911).] In: Bohr 2008, Vol. 1, 413–19.

Bohr, N. 1912. [The Rutherford Memorandum]. In: Bohr 2008, Vol. 2, 136–43.

Bohr, N. 1913a. On the Theory of Decrease of Velocity of Moving Electrified Particles on Passing through Matter. *Philosophical Magazine* 25: 10–31.

Bohr, N. 1913b. On the Constitution of Atoms and Molecules. *Philosophical Magazine* 26: 1–25, 476–502, 857–75.

Bohr, N. 1915. On the Quantum Theory of Radiation and the Structure of the Atom. *Philosophical Magazine* 30: 394–415.

Bohr, N. 1922a. On the Spectrum of Hydrogen. In: Bohr 1922b, 1–19.

Bohr, N. 1922b. *The Theory of Spectra and Atomic Constitution*. Cambridge: Cambridge University Press.

Bohr, N. 1925. Atomic Theory and Mechanics. *Nature* 116, No. 2927: 845–52.

Bohr, N. 1926. Atomtheorie und Mechanik. *Naturwissenschaften* 14: 1–10.

Bohr, N. 2008. *Collected Works*, Vol. 1–13. Amsterdam: Elsevier.

Born, M. 1926. *Problems of Atomic Dynamics*. Cambridge (MA): Massachusetts Institute of Technology.

Born, M. and Heisenberg, W. 1923. Die Elektronenbahnen im angeregten Heliumatom. *Zeitschrift für Physik* 16: 229–43.

Born, M. and Heisenberg, W. 1928a. Die Quantenmechanik. [German typescript of Born–Heisenberg 1928b; deposited in the Staattsbibliothek, Berlin, Germany].

Born, M. and Heisenberg, W. 1928b. La mécanique des quanta. In: *Électrons et photons – Rapports et discussions de cinquième Conseil de Physique, Institut International de Physique Solvay* (pp. 281–304). Paris: Gauthiers-Villars.

Born, M. and Heisenberg, W. 2009. Quantum Mechanics. In: *Quantum Theory at the Crossroads*, G. Bacciagalupi and A. Valentini, 407–40. Cambridge: Cambridge University Press. [Translated from Born–Heisenberg 1928a].

Born, M. and Jordan, P. 1925. Zur Quantenmechanik. *Zeitschrift für Physik* 34: 858–88.

Born, M. and Heisenberg, W. and Jordan, P. 1926. Zur Quantenmechanik II. *Zeitschrift für Physik* 35: 557–615.

Brackett, F. S. 1922. A New Series of Spectral Lines. *Nature* 109: 206.

Brakke, K. E. and Savage-Rumbaugh, E. S. 1995. The Development of Language Skills in Bonobo and Chimpanzee—I; Comprehension. *Language and Communication* 15: 121–48.

Brakke, K. E. and Savage-Rumbaugh, E. S. 1996. The Development of Language Skills in Bonobo and Chimpanzee—II; Production. *Language and Communication* 16: 361–80.

Brewer, W. F. and Lambert, B. L. 2001. The Theory-Ladenness of Observation and the Theory-Ladenness of the Rest of the Scientific Process. *Philosophy of Science* 68: S176–S186.

Bullinger, A. F. et al. 2013. Bonobos, *Pan Paniscus*, chimpanzees, *Pan troglodytes*, and marmosets, *Callithrix jacchus*, prefer to feed alone. *Animal Behavior* 85: 51–60.

Call, J. 2001. Chimpanzee Social Cognition. *Trends in Cognitive Science* 5: 388–93.

Call, J. 2009. Contrasting the Social Cognition of Humans and Nonhuman Apes: The Shared Intentionality Hypothesis. *Topics in Cognitive Science* 1: 368–79.

Call, J. and Tomasello, M. 1999. A Non-verbal False Belief Test. *Child Development* 70: 381–95.

Camillieri, K. 2009. *Heisenberg and the Interpretation of Quantum Mechanics*. Cambridge: Cambridge University Press.

Cardwell, D. S. L. 1967. Some Factors in the Early Development of the Concepts of Power, Work and Energy. *British Journal for the History of Science* 3: 209–24.

Cardwell, D. S. L. 1971. *From Watt to Clausius*. Ithaca (NY): Cornell University Press.

Carnap, R. 1932a. Überwindung der Metaphysik durch die logische Analyse der Sprache. *Erkenntnis* 2: 219–41.

Carnap, R. 1932b. Die physikalische Sprache als Universalsprache der Wissenschaft. *Erkenntnis* 2: 432–65.

Carnap, R. 1936/37. Testability and Meaning. *Philosophy of Science* 3: 420–71; 4: 2–40.

Carnap, R. 1939. Foundations of Logic and Mathematics. *International Encyclopedia of Unified Science* I, No. 3: 1–71.

Carnap, R. 1956. The Methodological Character of Theoretical Concepts. In: *Minnesota Studies in the Philosophy of Science*, Vol. 1, eds. H. Feigl and M. Scriven, 38–75. Minneapolis: University of Minnesota Press.

Carnap, R. 1958. Beobachtungssprache und Theoretische Sprache. *Dialectica* 12: 236–48.

Carnap, R. 1959/2000. Theoretical Concepts in Science. *Studies in the History and Philosophy of Science* 31: 158–72.

Carnot, S. 1824. *Réflexions sur la puissance motrice du feux.* Paris: Bachelier.

Carpenter, M. and Pennington, B. and Rogers, S. J. 2001. Understanding of Other's Intentions in Young Children with Autism and Developmental Delays. *Journal of Autism and Developmental Disorders* 31: 589–99.

Carpenter, M. and Pennington, B. and Rogers, S. J. 2002. Interrelations among Social Cognitive Skills in Young Children with Autism and Developmental Delays. *Journal of Autism and Developmental Disorders* 32: 91–106.

Carpenter, M. and Tomasello, M. and Striano, T. 2005. Role Reversal Imitation and Language in Typically Developing Infants and Children with Autism. *Infancy* 8: 253–78.

Carvalho, S. *et al.* 2008. Chaînes opératoires and the resource exploitation strategies in chimpanzee (*Pan troglodytes*) nut cracking in. *Journal of Human Evolution* 55: 148–63.

Cassidy, D. 2009. *Beyond Uncertainty.* New York: Bellevue Literary Press.

Chang, H. 2004. *Inventing Temperature.* Cambridge: Cambridge University Press.

Chang, H. and Yi, S. W. 2005. The Absolute and its Measurement. *Annals of Science* 62: 281–308.

Cleland, C. E. 2002. Methodological and Epistemic Differences between Historical and Experimental Science. *Philosophy of Science* 69: 474–96.

Cleland, C. E. 2011. Prediction and Explanation in Historical Natural Science. *British Journal for the Philosophy of Science* 72: 551–82.

Conway, A. W. 1907. On Series Spectra. *Scientific Proceedings of the Dublin Royal Society* 11: 181–83.

Coriolis, G. G. 1829. *Du calcul de l'effet des machines.* Paris: Carilian-Goeury.

Crick, J. *et al.* 2013. The roles of food quality and sex in chimpanzee sharing behavior (*Pan troglodytes*). *Behavior* 150: 1203–1224.

Cropper, W. H. 1987. Carnot's Function: Origins of the Thermodynamic Concept of Temperature. *American Journal of Physics* 55: 120–29.

Černík, V. 1978. Problém zákona v marxistickej metodológii vied. Bratislava: Vydavateľstvo Pravda.

Darden, L. 2002. Strategies for Discovering Mechanisms. *Philosophy of Science* 69: S354–S365.

Darden, L. 2008. Thinking Again about Biological Mechanisms. *Philosophy of Science* 75: 958–69.

Darrigol, O. 1992. *From c-Numbers to q-Numbers*. Berkeley: University of California Press.

Darrigol, O. 2012. *The History of Optics from Greek Antiquity to Nineteenth Century*. Oxford: Oxford University Press.

Davidson, D. 1999. The Development of Thought. *Erkenntnis* 51: 511–21.

Deblauwe, I. et al. 2006. Use of tool set by *Pan troglodytes troglodytes* to obtain termites (*Macrotermes*) in the periphery of the Dja Biosphere Reserve, Southeast Cameroon. *American Journal of Primatology* 68: 1191–1196.

De Luc, J.-A. 1784. *Recherches sur les modifications de l'atmosphere*, Tome II. Paris: Duchesne.

Desaguliers, J. T. 1744. *A Course of Experimental Philosophy*, Vol. II. London: Innys – Senex – Longman.

Dooley, P. C. 2005. *The Labour Theory of Value*. London: Routledge.

Ellis, B. 1960. Some Fundamental Problems of Direct Measurement. *Australasian Journal of Philosophy* 38: 37–47.

Evans, E. J. 1913. The Spectra of Helium and Hydrogen. *Nature* 92: 5.

Feather, N. 1968. *An Introduction to the Physics of Mass, Length and Time*. Chicago: Aldine.

Feigl, H. 1956. Some Major Issues and Developments in the Philosophy of Science of Logical Empiricism. In: *Minnesota Studies in the Philosophy of Science*, Vol. 1, eds. H. Feigl, H. and M. Scriven, 3–37. Minneapolis: University of Minnesota Press.

Feigl, H. 1970a. The "Orthodox" View of Theories. In: *Minnesota Studies in the Philosophy of Science*, Vol. 4, eds. M. Radner and S. Winokur, 3–15. Minneapolis: University of Minnesota Press.

Feigl, H. 1970b. Beyond Peaceful Coexistence. In: *Minnesota Studies in the Philosophy of Science*, Vol. 5, eds. R. H. Stuewer, 3–11. Minneapolis: University of Minnesota Press.

Feigl, H. 1974. Empiricism at Bay? In: *Methodological and Historical Essays on Natural and Social Science*, eds. R. S. Cohen and M. Wartofsky, 1–20. Dordrecht: Reidel.

Feyerabend, P. K. 1958. An Attempt at a Realistic Interpretation of Experience. *Proceedings of the Aristotelian Society* 58: 143–70.

Feyerabend, P. K. 1962. Explanation, Reduction, and Empiricism. In: *Minnesota Studies in the Philosophy of Science*, Vol. 3, eds. H. Feigl and G. Maxwell, 28–97. Minneapolis: University of Minnesota Press.

Feyerabend, P. K. 1965. Problems of Empiricism. In: *Beyond the Edge of Certainty*, ed. R. G. Colodny, 145–260. Englewood Cliffs, NJ: Prentice Hall.

Feyerabend, P. K. 1970. Against Method. In: *Minnesota Studies in the Philosophy of Science*, Vol. 4, eds. M. Radner and S. Winokur, 17–129. Minneapolis: University of Minnesota Press.

Fields, W. M. and Sagerdahl, P. and Savage-Rumbaugh, E. S. 2007. The Material Practices of Ape Language Research. In: *The Cambridge Handbook of Sociocultural Anthropology*, eds. J. Valsinen and A. Rosa, 164–86. Cambridge: Cambridge University Press.

Fisher, S. E. 2006. Tangled Webs. *Cognition* 101: 270–97.

Dulong, P.-L. 1817. Recherches sur la mesure de températures et sur les lois de la communication de la chaleur. *Annales de chimie et de physique* 7: 113–54, 225–64, 337–67.

Dulong, P.-L. and Petit, A.-T. 1816. Recherches sur les lois de dilation des solides, des liquides et des fluides élastiques, et sur la mesure de températures. *Annales de chimie et de physique* 2: 240–53.

Forrester, J. 1975. Chemistry and the Conservation of Energy. *Studies in the History and Philosophy of Science* 6: 273–313.

Fowler, A. and Sommer, V. 2006. Subsistence technology of Nigerian chimpanzees. *International Journal of Primatology* 28: 997–1023.

Franck, J. and Hertz, G. 1914. Über Zusammenstöße zwischen Elektronen und Molekülen des Quecksilberdampfes und die Ionisierungsspannung desselben. *Verhandlungen der deutschen physikalischen Gesellschaft* 16: 457–67.

Franck, J. and Hertz, G. 1919. Die Bestätigung der Bohrschen Atomtheorie im optischen Spektrum durch Untersuchungen der unelastischen Zusammenstöße langsamer Elektronen mit Gasmolekülen. *Physikalische Zeitschrift* 20: 132–43.

Frank, P. 1917. Die Bedeutung der physikalischen Erkenntnistheorie Machs für das Geistesleben der Gegenwart. *Naturwissenschaften* 5: 65–72.

Frank, P. 1928. Über die „Anschaulichkeit" der physikalischen Theorien. *Naturwissenschaften* 16: 121–28.

Gardner, R. A. and Gardner, B. T. 1969. Teaching Sign Language to a Chimpanzee. *Science* 165, No. 3894: 664–72.

Gardner, B. T. and Gardner, R. A. 1971. Two-way Communication with a Chimpanzee. In: *Behavior of nonhuman primates*, Vol. 4, A. Schreir and F. Stollnitz, 117–84. New York: Academic Press.

Gay-Lussac, J.-L. 1802a. Recherches sur la dilatation des gaz et des vapeurs. *Annales de Chimie* 43: 137–75.

Gay-Lussac, J.-L. 1802b. Enquiries Concerning the Dilatation of the Gases and Vapors. *Journal of Natural Philosophy, Chemistry and Arts* 3: 207–16, 257–67.

Geiger, H. and Marsden, E. 1909. On a Diffuse Reflection of the α-Particles. *Proceedings of the Royal Society of London* (Series A) 82: 495–500.

Gilby, I. C. 2006. Meat sharing among the Gombe chimpanzees. *Animal Behavior* 71: 953–63.

Gilby, I. C. et al. 2010. No evidence of short-term exchange of meat-for-sex among chimpanzees. *Journal of Human Evolution* 59: 44–53.

Gill, T. V. 1977. Conversations with Lana. In: Rumbaugh, ed. 1977: 225–46.

Gill, T. V. and Rumbaugh, D. M. 1977. Training Strategies and Tactics. In: Rumbaugh, ed. 1977: 157–62.

Glennan, S. S. 2010. Ephemeral Mechanisms and Historical Explanation. *Erkenntnis* 72: 251–66.

Gomes, C. M. and Boesch, C. 2009. Wild chimpanzees exchange meat for sex on long-term basis. *PLoS ONE* 4: e5116.

Gubrium, J. F. and Holstein, J. A. 2008. Narrative Etnography. In: *Emergent Methods*, eds. S. Hesse-Biber and P. Leavy, 241–64. New York, Guilford.

Habermas, J. 1981. *Theorie des kommunikativen Handelns*, Bd. 1. Frankfurt am Main: Suhrkamp.

Habermas, J. 1999a. Einleitung: Realismus nach der sprachpragmatischen Wende. In: Habermas 1999b: 7–64.

Habermas, J. 1999b. *Wahrheit und Rechtfertigung*. Frankfurt am Main: Suhrkamp.

Habermas, J. 2001. *On the Pragmatics of Social Interaction*. Cambridge (MA): MIT Press.

Habermas, J. 2005a. Introduction: Realism after the Linguistic Turn. In: Habermas 2005b: 1–50.

Habermas, J. 2005b. *Truth and Justification*. Cambridge (MA): MIT Press.

Habermas, J. 2009. Laudatio für Michael Tomasello, gehalten anlässlich der Verleihung des Hegel-Preises am 16. Dezember 2009 in Stuttgart. [Accessible on the url-adress http://www.stuttgart.de/img/mdb/item/383875/51478.pdf].

Halas, J. (2013): *Metodológia kritickej sociálnej vedy*. Bratislava: Vydavateľstvo Univerzity Komenského.

Halas, J. (2015): Abstrakcia a idealizácia ako metódy sociálno-humanitných disciplín. *Organon F* 22: 71–89.

Halley, E. 1693. An Account of Several Experiments Made to Examine the Nature of the Expansion and Contraction of Fluids by Heat and Cold, in order to Ascertain the Divisions of the Thermometer, and to Make that Instrument, in all Places, without Adjusting by a Standard. *Philosophical Transactions of the Royal Society of London* 17: 650–56.

Hanus, D. and Call, J. 2004. Quantity Based Judgments by Orangutans, Gorillas, and Bonobos. *Folia Primatologica* 75: 381–82.

Hanus, D. and Call, J. 2007. Discrete Quantity Judgments in the Great Apes. *Journal of Comparative Psychology* 121: 241–49.

Hanus, D. and Call, J. 2008. Chimpanzees Infer the Location of a Reward Based on the Effect of Weight. *Current Biology* 18: R370–R372.

Hanus, D. and Call, J. 2009. Great Apes' Capacities to Recognize Relational Similarities. *Cognition* 110: 147–54.

Hanus, D. *et al.* 2011. Comparing the Performances of Apes (*Gorilla gorilla, Pan troglodytes, Pongo pygmeus*) and Human Children (*Homo sapiens*) in the Floating Peanut Task. *PLoS One* 6: 1–13.

Hanzel, I. 1999. *The Concept of Scientific Law in the Philosophy of Science and Epistemology*. Dordrecht: Kluwer.

Hanzel, I. 2010. Mistranslations of "Schein" and "Erscheinung". *Science and Society* 74: 509–37.

Hanzel, I. 2014. 'The Circular Course of Our Representation'. In: *Marx's Capital and Hegel's Logic*, eds. F. Moseley and T. Smith, 214–39. Leiden: Brill.

Hanzel, I. and Černík, V. and Viceník, J. 1994. What is a Category? *Metaphilosophy* 25: 181–93.

Hare, B. and Call, J. and Tomasello, M. 2000. Chimpanzees Know What Conspecifics Do and Do Not See. *Animal Behavior* 59: 771–85.

Hayashi M. and Y. Mizuno and T. Matsuzawa. 2005. How does stone-tool use emerge. *Primates* 46: 91–102.

Hayashi M. and H. Takeshita and T. Matsuzawa. 2006. Cognitive development in apes and humans assessed by object manipulation. In: Matsuzawa and Tomonaga and Tanaka, eds. 2006: 395–410.

Hayes, K. J. and Hayes, C. 1951: The Intellectual Development of a Home-Raised Chimpanzee. *Proceedings of the American Philosophical Society* 95: 105–09.

Hegel, G. W. F. 1986. *Wissenschaft der Logik*, Bd. I, II. Frankfurt am Main: Suhrkamp. Press.

Heilbron, J. L. 1968. The Scattering of α and β Particles and the Rutherford Atom. *Archive for the History of Exact Sciences* 4: 247–307.

Heilbron, J. L. and Kuhn, T. S. 1969. The Genesis of the Bohr Atom. *Historical Studies in the Physical Sciences* 1: 211–90.

Heisenberg, W. 1925. Über quantentheoretische Umdeutung kinematischer und mechanischer Beziehungen. *Zeitschrift für Physik* 33: 879–93.

Heisenberg, W. 1926. Über quantentheoretische Kinematik und Mechanik. *Mathematische Annalen* 95: 683–705.

Heisenberg, W. 1927. Über den anschaulichen Inhalt der quantentheoretischen Kinematik und Mechanik. *Zeitschrift für Physik* 43: 172–98.

Heisenberg, W. 1928/1984. Erkenntnistheoretische Probleme in der modernen Physik. Published in Heisenberg, W.: *Gesammelte Werke*, Abteilung C, Bd. I (pp. 22–28). München: Piper.

Heisenberg, W. 1934. Wandlungen in den Grundlagen der exakten Naturwissenschaften in der jüngsten Zeit. *Naturwissenschaften* 22: 697–702.

Heisenberg, W. 1969. *Das Teil und das Ganze*. München: Piper.

Heisenberg, W. et al. 1931/1932. Diskussion über Kausalität und Quantenmechanik. *Erkenntnis* 2: 183–88.

Hempel, C. G. 1951. The Concept of Cognitive Significance. *Proceedings of the American Academy of Arts and Sciences* 80: 61–77.

Hempel, C. G. 1952. Fundamentals of Concept Formation. *International Encyclopedia of Unified Science*, Vol. II: 1–93.

Hempel, C. G. 1958. The Theoretician's Dilemma. In: *Minnesota Studies in the Philosophy of Science*, Vol. 2, eds. H. Feigl and M. Scriven, 37–98. Minneapolis, University of Minnesota Press.

Hempel, C. G. 1970. On the Standard Conception of Scientific Theories. In: *Minnesota Studies in the Philosophy of Science*, Vol. 4, eds. M. Radner and S. Winokur, 142–63. Minneapolis: University of Minnesota Press.

Henry, J. E. 2000. Adam Smith and the Theory of Value. *History of Economics Review* 31: 1–13.

Herrmann, E. et al. 2010. Differences in the Cognitive Skills of Bonobos and Chimpanzees. *PLoS One* 5: e12348.

Hill, K. 2002. Altruistic Cooperation during Foraging by the Ache, and the Evolved Human Predisposition to Cooperate. *Human Nature* 13: 105–28.

Hill, K. and Hurtado, A. M. 1996. *Ache Life History*. New York: Aldine de Gruyter.

Hindriks, F. 2013. Explanation, Understanding, and Unrealistic Models. *Studies in the History and Philosophy of Science* 44: 523–31.

Hitchcock, C. and Woodward, J. 2003. Explanatory Generalizations, Part II. *Noûs* 37: 181–99.

Hockings, K. J. *et al.* 2007. Chimpanzees share forbidden fruits. *PLoS ONE* 2: e886.

Horner, A. *et al.* 2011. Spontaneous prosocial choice by chimpanzees. *Proceedings of the National Academy of Sciences of the United States of America* 108: 13847–13851.

Hueckel, G. 2000. On the 'Insurmountable Difficulties, Obscurity, and Embarrassment' of Smith's Fifth Chapter. *History of Political Economy* 32: 317–45.

Huggins, W. 1880. Mr. Huggins's Observatory, Upper Tulse Hill. *Monthly Notices of the Royal Astronomical Society* 40: 227–28.

Huygens, C. 1673. *Horologium Oscillotarium*. Parisiis: F. Muguet.

Huygens, C. 1986. *The Pendulum Clock*. Ames: Iowa State University Press.

Imbert, C. 2013. Relevance, Not Invariance, Explanatoriness, Not Manipulability. *Philosophy of Science* 80: 625–36.

Jaeck, H.-P. 1988. *Genesis und Notwendigkeit*. Berlin: Akademie-Verlag.

Jaffares, B. 2010. Guessing the Future of the Past. *Biology and Philosophy* 25: 125–42.

Jantzen, W. 2003. Zum Verhältnis von Ideellem und Ideal bei Il'enkov und Leont'ev. In: *Ein Diamant schleift den anderen*, eds. W. Jantzen and B. Siebert, 316–40. Berlin: Lehmanns Media.

Jensen, K. *et al.* 2006. What's in for me? *Proceedings of the Royal Society B* 273: 1013–1021.

Jones, N. 2013. Don't Blame Idealizations. *Journal for the General Philosophy of Science* 44: 85–100.

Joule, J. P. 1843. On the Caloric Effects of Magnetoelectricity and the Mechanical Value of Heat. *Philosophical Magazine* 23: 263–76, 347–55, 435–43.

Joule, J. P. 1847. On Matter, Living Force and Heat. *Manchester Courier*, May 5 and 12. Reprinted in *Scientific Papers*, J. P. Joule, 265–76. London 1884: Taylor and Francis.

Joule, J. P. 1850. On the Mechanical Equivalent of Heat. *Philosophical Transactions of the Royal Society of London* 140: 61–82.

Joule, J. P. and Thomson, W. 1854. On the Thermal Effects of Fluids in Motion, Part 2. *Philosophical Transactions of the Royal Society of London* 144: 321–64.

Kant, I. 1938. *Gesammelte Schriften*, Bd. XXII. Berlin: Walter de Gruyter und Co.

Kant, I. 1973. *Critique of Pure Reason*. London: MacMillan.

Kant, I. 1993. *Opus Postumum*. Cambridge: Cambridge University Press.

Kant, I. 1998. *Critique of Pure Reason*. Cambridge: Cambridge University Press.

Kaushil, S. 1973. The Case of Adam Smith's Value Analysis. *Oxford Economic Papers* 25: 60–71.

Kitahara-Frisch, J. 1993. The origin of secondary tools. In: *The use of tools by human and nonhuman primates*, eds. A. Bethelet and J. Chavaillon, 239–46. Oxford: Clarendon.

Koťátko, P. 2005. Jazyk a svět. In: *Jazyk – Logika – Věda*, ed. P. Sousedík, 11–38. Praha: Filosofie.

Krachun, C. et al. 2009. A competitive nonverbal false belief task for children and apes. *Developmental Science* 12: 521–35.

Krachun, C. et al. 2010. A new change-of-contents false belief test. *International Journal of Comparative Psychology* 23: 145–65.

Kramers, H. A. 1923. Über das Modell des Heliumatoms. *Zeitschrift für Physik* 13: 312–42.

Krüger, H.-P. 1999. *Zwischen Weinen und Lachen*. Berlin: Akademie Verlag.

Krüger, H.-P. 2010. *Gehirn, Verhalten und Zeit*. Berlin: Akademie Verlag.

Lanczos, C. 1926. Über eine feldmäßige Darstellung der neuen Quantmechanik. *Zeitschrift für Physik* 35: 812–30.

Landauer, J. 1898. *Spectrum Analysis*. New York: John Wiley and Sons.

Lavoisier, A. L. and Laplace, P. S. 1784. Mémoire sur la chaleur. *Mémoires de l'Académie des sciences*, année 1780: 283–333.

Liebal, K. et al. 2008. Helping and Cooperation in Children with Autism. *Journal of Autism and Developmental Disorders* 38: 224–38.

Liszkowski, U. 2004. 12-Month-Olds Point to Share Attention and Interest. *Developmental Science* 7: 297–307.

Liszkowski, U. 2005. Human Twelve-Month-Olds Point Cooperatively to Share Interest with and Helpfully Provide Information for a Communicative Partner. *Gesture* 5 (1–2): 135–54.

Liszkowski, U. and Carpenter, M. and Tomasello, M. 2007a. Pointing Out New News, Old News, and Absent Referents at 12 Months of Age. *Developmental Science* 10: F1–F7.

Liszkowski, U. and Carpenter, M. and Tomasello, M. 2007b. Reference and Attitude in Infant Pointing. *Journal of Child Language* 34: 1–20.

Liszkowski, U. et al. 2006. 12- and 18-Month Olds Point to Provide Information to Other. *Journal of Cognition and Development* 7: 173–87.

Liszkowski, U. et al. 2012. A Prelinguistic Gestural Universal of Human Communication. *Cognitive Science* 36: 698–713.

Lockyer, N. and Chisholm-Batten, A. W. and Peddler, A. 1901. Total Eclipse of the Sun, January 22, 1898. *Philosophical Transactions of the Royal Society A* 197: 151–227.

Lyman, T. 1914. An Extension of the Spectrum in the Extreme-Violet. *Physical Review* 3: 504–05.

Machamer, P. 2004. Activities and Causation. *International Studies in the Philosophy of Science* 18: 27–39.

Machamer, P. and Darden, L. and Craver, C. F. 2000. Thinking about Mechanism. *Philosophy of Science* 67: 1–25.

[machine à diviser] Description of this instrument can be found under the following url-address: http://www.louislegrand.eu/index.php?option=com_content&view=article&id=156:mes2-2&catid=11:appareils-de-mesure&Itemid=10.

Marx, K. 1976. *Ökonomische Manuskripte 1857–1858*, Teil 1. MEGA II/1.1. Berlin: Dietz.

Marx, K. [1858] 1980a. *Zur Kritik der politischen Ökonomie*, Erstes Heft. Franz Duncker: Berlin. MEGA II/2. Berlin, Dietz: 99–255.

Marx, K. 1980b. *Zur Kritik der politischen Ökonomie (Manuskript 1861–1863)*, Teil 5. MEGA II/3.5. Berlin, Dietz.

Marx, K. [1867] 1983. *Das Kapital. Erster Band.* Hamburg: Otto Meissner. MEGA II/5. Berlin: Dietz.

Marx, K. [1872] 1987. *Das Kapital. Erster Band.* Hamburg 1872: Otto Meissner. MEGA II/6. Berlin: Dietz.

Marx, K. 1988a. Sechstes Kapitel. Resultate des unmittelbaren Produktionsprocesses. In: Marx 1988c, 24–135.

Marx, K. 1988b. Das Kapital (Ökonomisches Manuskript 1863–1865). Zweites Buch (Manuskript I). In: Marx 1988c, 137–381.

Marx, K. 1988c. *Das Kapital. Ökonomische Manuskripte 1863–1867*, Teil 1. MEGA II/4.1. Berlin: Dietz.

Marx, K. [1872] 1989. *Le Capital*. Paris: Maurice Lachatre et Cie. MEGA II/7. Berlin: Dietz.

Marx, K. 1992. *Ökonomische Manuskripte 1863–1867*, Teil 2 *(Manuskript 1863/65 zum 3. Buch des „Kapital")*. MEGA II/4.2. Berlin – Amsterdam: Dietz – Internationales Institut für Sozialgeschichte.

Marx, K. 2003a. Manuskripte zum dritten Buch des Kapitals, 1871–1882. Pp. 3–163 in Marx, K. and Engels, F. 2003a.

Marx, K. 2003b. Value, Price and Profit. Pp. 141–186 in MEGA I/20. Berlin: Akademie Verlag.

Marx, K. 2008. *Manuskripte zum zweiten Buch des „Kapitals"1868–1881*. MEGA II/11. Berlin: Akademie Verlag.

Marx, K 2012a. Zweites Buch. Der Cirkulationsprozess des Kapitals. Pp. 32–44 in Marx 2012d.

Marx, K. 2012b. Zweites Buch. Der Cirkulationsprozess des Kapitals (Manuskript IV). Pp. 285–363 in Marx 2012d.

Marx, K. 2012c. Drittes Buch. Pp. 7–30, 57–283 in Marx 2012d.

Marx, K. 2012d. *Ökonomische Manuskripte 1863–1868*, Teil 3 *(Manuskripte zum zweiten und dritten Buch des „Kapitals" 1867/1868)*. MEGA II/4.3. Berlin: Akademie Verlag.

Marx, K. and Engels, F. 2003a. Manuskripte und redaktionelle Texte zum dritten Buch des „Kapitals", 1871 bis 1895. MEGA II/14. Berlin: Akademie Verlag.

Marx, K. and Engels, F. 2003b. *Briefwechsel*. Januar 1858 bis August 1859. MEGA III/9. Berlin: Akademie Verlag.

Marzke, M. W. et al. 1998: EMG Study of Hand Muscle Recruitment during Hard Hammer Percussion Manufacture of Oldowan Tools. *American Journal of Physical Anthropology* 105: 315–22.

Matsuzawa, T. 1996, Chimpanzee intelligence in nature and captivity. In: *Great Ape Societies*, eds. W. C. McGrew and L. P. Marchant and T. Nishida, 196–212. Cambridge: Cambridge University Press.

Matsuzawa, T. 2001. Primate foundations of human intelligence. In: *Primate Origins of Human Cognition and Behavior*, ed. T. Matsuzawa, 3–25. Tokio: Springer.

Matsuzawa, T. and Humle, T. and Suguyami, Y. eds. 2011. *The chimpanzees of Bossou and Nimba*. Dordrecht: Springer.

Matsuzawa, T. and Tomonaga, M. and Tanaka, M. eds. 2006. *Cognitive development in chimpanzees*. Dordrecht: Springer.

Maury, A. C. and Pickering, E. C. 1897. Spectra of Bright Stars. *Annals of the Harvard College Observatory* 28: 1–128.

McAllister, J. W. 2011. What do Patterns in Empirical Data tell us about the Structure of the World? *Synthese* 182: 73–87.

McGucken, W. 1969. *Nineteenth-Century Spectroscopy*. Baltimore: Johns Hopkins University Press.

McKie, D. and Heathcote, N. H. de V. 1975. *The Discovery of Specific and Latent Heats*. New York: Arno Press.

McLennan, M. R. 2011. Tool-use to obtain honey by chimpanzees at Bulundi: new record from Uganda. *Primates* 52: 315–22.

McMullin, E. 1985. Galilean Idealization. *Studies in the History and Philosophy of Science* 16: 247–73.

Mehmet, E. 2008. Theory-Laden Observation and Incommensurability. *Organon F* 15: 3–19.

Melis, A. P. and Hare, B. and Tomasello, M. 2006a. Engineering Cooperation in Chimpanzees. *Animal Behavior* 72: 275–86.

Melis, A. P. and Hare, B. and Tomasello, M. 2006b. Chimpanzees Recruit the Best Collaborators. *Science* 31, No. 5765: 1297–1300.

Mill, J. S. [1848] 1965. *Principles of Political Economy*, Vol. II. London: Routledge and Kegan Paul.

Mitani, J. C. and Watts, D. P. 2001. Why do chimpanzees hunt and share meat? *Animal Behavior* 61: 915–24.

Mitani, J. C. and Merriwether, D. A. and Zhang, C. 2000. Male affiliation, cooperation and kinship in wild chimpanzees. *Animal Behavior* 59: 885–93.

Naldi, N. 2013. Adam Smith on Value and Prices. In: *The Oxford Handbook of Adam Smith*, eds. C. J. Berry and M. Paganelli and C. Smith, 290–306. Oxford: Oxford University Press.

Nowak, L. 1972. Laws of Science, Theories, Measurement. *Philosophy of Science* 39: 533–48.

Nowak, L. 1980. *The Structure of Idealization*. Dordrecht: Reidel.

Nowak, L. and Nowakowa, I. 2000. *Idealization X*. Amsterdam, Rodopi.

O'Donnell, R. 1990. *Adam Smith's Theory of Value and Distribution*. London: MacMillan.

Pack, S. J. 2010. *Aristotle, Adam Smith and Karl Marx*. Cheltenham: Edward Elgar.

Paschen, F. 1908. Zur Kenntnis ultraroter Linienspektren. I. *Annalen der Physik* 27: 537–70.

Parker, S. T. and Gibson, K. R. 1977. Object manipulation, tool use, and sensorimotor intelligence as feeding adaptations in Cebus monkeys and Great Apes. *Journal of Human Evolution* 6: 623–41.

Pascual-Garrido, A. et al. 2012. Obtaining raw material: plants as tool sources for Nigerian chimpanzees. *Folia Primatologica* 83: 24–44.

Pauli, W. 1919. Merkurperihelbewegung und Strahlungablenkung in Weyls Gravitationstheorie. *Verhandlungen der Deutschen Physikalischen Gesellschaft* 21: 742–50.

Pauli, W. 1979. *Wissenschaftlicher Briefwechsel mit Bohr, Einstein, Heisenberg u. a.* Band 1. New York: Julius Springer.

Pawson, R. 1989. *A Measure for Measure*. London: Routledge.

Petit, A. T. 1818. Sur l'Emploi du principe des forces vives dans le calcul de l'effet des machines. *Annales de chimie et de physique* 8: 287–302.

Pfund, A. H. 1924. The Emission of Nitrogen and Hydrogen in the Infrared. *Journal of the Optical Society of America* 9: 193–96.

Pickering, E. C. 1896a. Stars having Peculiar Spectra. *Astronomische Nachrichten* 141: 169.

Pickering, E. C. 1896b. Stars having Peculiar Spectra. *Astrophysical Journal* 4: 369–70.

Pickering, E. C. 1897a. The Spectrum of ζ Puppis. *Astrophysical Journal* 5: 92–4.

Pickering, E. C. 1897b. The Spectrum of ζ Puppis. *Astrophysical Journal* 6: 259.

Pika, S. and Zuberbühler, K. 2008. Social Games between Bonobos and Humans. *American Journal of Primatology* 70: 207–10.

Premack, D. and Woodruff, G. 1978a. Does the Chimpanzee Have a Theory of Mind? *Behavioral and Brain Sciences* 4: 515–28.

Premack, D. and Woodruff, G. 1978b. Chimpanzee Problem-Solving. *Science* 202, No. 4367: 532–35.

Pruetz, J. D. and Bertolani, P. 2007. Savanna chimpanzees, *Pan troglodytes verus*, hunt with tools. *Current Biology* 17: 412–17.

Pruetz, J. D. and Lindshield, S. 2012. Plant-food and tool transfer among savanna chimpanzee at Fongoli, Senegal. *Primates* 53: 133–45.

Rochefort-Maranda, G. 2008. The Double-Language Model and the Theory-Ladenness of Our Observation Reports. In: *Volume of Abstracts; Sixth European Congress of Analytic Philosophy*, eds. V. Kukushkina, and L. Kijania-Placek, 320. Krakow, Jiagellonian University.

Rakoczy, H. 2008. Du, Ich, Wir. In: *Other Minds*, ed. R. Schubotz, 93–109. Paderborn: Mentis.

Rescher, N. 1981. On the Status of the "Things in Themselves" in Kant. *Synthese* 47: 289–99.

Richmann, G. W. 1747–1748. De quantitate caloris. *Novi commentari academicae scientiarum imperialis petropolitanae* 1: 152–67.

Ritz, W. 1903. Zur Theorie der Serienspektren. *Annalen der Physik* 12: 264–310.

Ritz, W. 1908a. Magnetische Atomfelder und Serienspektren. *Annalen der Physik* 25: 660–96.

Ritz, W. 1908b. Über ein neues Gesetz der Serienspektren. *Physikalische Zeitschrift* 9: 521–29. Reprinted in Ritz 1911: 141–62.

Ritz, W. 1911. *Gesammelte Werke*. Paris: Gauthier-Villars.

Robinson, J. 1950. [Review of] *Karl Marx and the Close of his System*. *Economic Journal* 60: 358–63.

Robinson, J. 1964. *Economic Philosophy*. Harmondsworth: Penguin.

Robinson, J. 1966. *An Essay on Marx*. London: Macmillan.

Robinson, J. 1979. Foreword. Pp. vii–xv in Karl Kühne: *Economics and Marx*, Vol. I. New York, St. Martin's Press.

Rohwer, Y. and Rice, C. 2013. Hypothetical Pattern Idealization and Explanatory Models. *Philosophy of Science* 80: 334–55.

Roncaglia, A. 2006. *The Wealth of Ideas*. Cambridge: Cambridge University Press.

Rumbaugh, D. M. 1993. Primate Language and Cognition. *Social Research* 62: 711–30.

Rumbaugh, D. M. ed. 1977. *Language Learning by a Chimpanzee*. New York: Academic Press.

Rumbaugh, D. M. *et al*. 1973. Computer-Controlled Language Training System for Investigating the Language Skills of Young Apes. *Behavior Research Methods and Instrumentation* 5: 385–92.

Rumbaugh, D. M. *et al*. 1975. Conversations with a Chimpanzee in a Computer-Controlled Environment. *Biological Psychiatry* 10: 627–41.

Rumbaugh, D. M. and Gill, T. V. 1976a. Language and the Acquisition of Language-Type Skills by a Chimpanzee (*Pan*). *Annals of the New York Academy of Sciences* 270: 90–123.

Rumbaugh, D. M. and Gill, T. V. 1976b. The Mastery of Language-Type Skills by a Chimpanzee (*Pan*). *Annals of the New York Academy of Sciences* 280: 562–78.

Rumbaugh, D. M. and Gill, T. V. 1977. Lana's Acquisition of Language Skills. In: Rumbaugh, ed. 1977: 165–92.

Rumbaugh, D. M. and Washburn, D. A. 2003. *Intelligence of Apes and Other Rational Beings*. New Haven: Yale University Press.

Rumbaugh, D. M. and Gill, T. V. and Glasersfeld von, E. C. 1973. Reading and Sentence Completion by a Chimpanzee. *Science* 182, No. 4113: 731–33.

Rumbaugh, D. M. and Savage-Rumbaugh, E. S. 1978. Chimpanzee Language Research. *Behavior Research Methods and Instrumentation* 10: 119–31.

Rumbaugh D. M. and Savage-Rumbaugh S. E. and Hegel, M. T. 1988. Addendum to "Summation in Chimpanzee Pan troglodytes." *Journal of Experimental Psychology – Animal Behavioral Process* 14: 118–20.

Rumbaugh D. M. and Savage-Rumbaugh S. E. and Pate, J. L. 1987. Summation in Chimpanzee (*Pan troglodytes*). *Journal of Experimental Psychology – Animal Behavioral Process* 13: 107–15.

Rumbaugh, D. M. and Savage-Rumbaugh, E. S. and Washburn, D. A. 1996. Toward a New Look on Primate Learning and Behavior. *Japanese Psychological Research* 38: 113–25.

Rutherford, E. 1911. The Scattering of α– and β–particles and the Structure of the Atom. *Philosophical Magazine* 21: 669–88.

Rydberg, J. R. 1890a. Recherches sur la constitution des spectres d'émission des éléments chimiques. *Kungliga Svenska Vetenskaps-Akademiens Handlinger* 23: 1–151.

Rydberg, J. R. 1890b. Über den Bau der Linienspektren der chemischen Grundstoffe. *Zeitschrift für physikalische Chemie* 5: 227–32.

Rydberg, J. R. 1890c. Sur la constitution des spectres linéaires des éléments chimiques. *Comptes Rendus* 110: 394–97.

Rydberg, J. R. 1890d. On the Structure of Line-Spectra of Chemical Elements. *Philosophical Magazine* 29: 331–37.

Rydberg, J. R. 1896. Die neuen Grundstoffe des Cleveïtgases. *Annalen der Physik und Chemie* 58: 674–79.

Rydberg, J. R. 1897. The New Series in the Spectrum of Hydrogen. *Astrophysical Journal* 6: 233–38.

Sanz, C. M. and Morgan, D. B. 2007. Chimpanzee tool technology in the Goualougo traingle, Republic of Congo. *Journal of Human Evolution* 52: 420–33.

Sanz, C. M. and Morgan, D. B. 2009. Flexible and persistent tool-using strategies in honey gathering by wild chimpanzees. *International Journal of Primatology* 30: 411–27.

Sanz, C. M. and Morgan, D. B. 2010. The complexity of chimpanzee tool-use behaviors. In: *Chimpanzee Minds*, eds. E. V. Lonsdorf and S. S. Ross and T. Matsuzawa, 127–40. Chicago: Chicago University Press.

Sanz, C. M. and Morgan, D. B. 2011. Elemental variation in termite fishing of wild chimpanzees (*Pan troglodytes*). *Biology Letters* 7: 634–37.

Sanz, C. M. and Call, J. and Morgan, D. B. 2009. Design complexity in termite-fishing tools of chimpanzees (*Pan troglodytes*). *Biology Letters* 5: 293–96.

Sanz, C. M. and Morgan, D. B. and Gulick, S. 2004. New insights into chimpanzee tools and termites from the Congo basin. *American Naturalist* 64: 567–81.

Sanz, C. M. and Schoening, C. and, Morgan, D. B. 2009. Chimpanzees prey on army ants with specialized tool set. *American Journal of Primatology* 72: 17–24.

Saatsi, J. and Pexton, M. 2013. Reassessing Woodward's Account of Explanation. *Philosophy of Science* 80: 613–24.

Savage-Rumbaugh, E. S. 1981. Can Apes Use Symbols to Represent their World? *Annals of the New York Academy of Sciences* 364: 35–59.

Savage-Rumbaugh, E. S. 1984. Verbal Behavior at the Procedural Level in the Chimpanzee. *Journal of Experimental Analysis of Behavior* 41: 223–50.

Savage-Rumbaugh, E. S. 1986. *Ape Language*. New York: Columbia University Press.

Savage-Rumbaugh, E. S. 1987a. Communication, Symbolic Communication, and Language. *Journal of Experimental Psychology: General* 116: 288–92.

Savage-Rumbaugh, E. S. 1987b. A New Look at Ape Language. *Nebraska Symposium on Motivation* 35: 201–55.

Savage-Rumbaugh, E. S. 1990. Language as a Cause-Effect Communication System. *Philosophical Psychology* 3: 57–76.

Savage-Rumbaugh, E. S. 1993. Language Learnability in Man, Ape, and Dolphin. In: *Language and Communication*, eds. H. L. Roitblat and M. L. Herman and P. E. Nachtigall, 457–84. Hillsdale (NJ): Lawrence Erlbaum.

Savage-Rumbaugh, E. S. 1997. Why Are We Afraid of Apes with Language? In: *The Origins and Evolution of Intelligence*, eds. A. B. Scheibel and J. W. Schopf, 43–69. Sudbury (MA): Jones and Bartlett Production.

Savage-Rumbaugh, E. S. 1998. Scientific Schizophrenia with Respect to the Language Act. In: *Piaget, Evolution and Development*, eds. J. Langer and M. Killen, 145–70. Hillsdale (NJ): Lawrence Erlbaum.

Savage-Rumbaugh, E. S. 1999. Ape Language. In: *The Origins of Language*, ed. J. B. King, 115–88. Santa Fe: School of American Research Press.

Savage-Rumbaugh E. S. and Brakke, K. E. 1990. Animal Language. In: *Interpretation and Explanation in the Study of Animal Behavior*, Vol. 1, eds. M. Bekoff and D. Jamieson, 313–43. Boulder (CO): Westview Press.

Savage-Rumbaugh, E. S. and Fields, W. M. 2006. Rules and Tools. In: *The Oldowan*, eds. N. Toth and K. Schick, 223–42. Bloomington (IN): The Stone Age Institute.

Savage-Rumbaugh, E. S. and Hopkins, K. E. 1986. Awareness, Intentionality, and Acquired Communicative Behavior. In: *Dolphin Cognition and Behavior*, eds. R. J. Schusterman, and J. A. Thomas and F. G. Wood, 303–13. Hillsdale (NJ): Lawrence Erlbaum.

Savage-Rumbaugh, E. S. and Rumbaugh, D. M. 1977. Awareness, Intentionality, and Acquired Communicative Behavior. In: Rumbaugh, ed. 1977: 287–309.

Savage-Rumbaugh, E. S. and Rumbaugh, D. M. 1978. Symbolization, Language and Chimpanzees. *Brain and Language* 6: 265–300.

Savage-Rumbaugh, E. S. and Rumbaugh, D. M. 1980. Requisites of Symbolic Communication. *Psychological Record* 30: 305–15.

Savage-Rumbaugh, E. S. and Fields, W. M. and Spircu, T. 2004. The Emergence of Knapping and Vocal Expression Embedded in *Pan/Homo* Culture. *Biology and Philosophy* 19: 541–75.

Savage-Rumbaugh, S. and Fields, W. M. and Taglialatela, J. P. 2000. Ape Consciousness – Human Consciousness. *American Zoologist* 40: 910–21.

Savage-Rumbaugh, E. S. and Rumbaugh, D. M. and Boysen, S. 1978a. Symbolic Communication between Two Chimpanzees (*Pan Troglodytes*). *Science* 201, No. 4356: 641–44.

Savage-Rumbaugh, E. S. and Rumbaugh, D. M. and Boysen, S. 1978b. Linguistically Mediated Tool Use and Exchange by Chimpanzees (*Pan Troglodytes*). *Behavioral and Brain Sciences* 1: 539–54.

Savage-Rumbaugh, E. S. and Rumbaugh, D. M. and Boysen, S. 1980, Do Apes Use Language? *American Scientist* 68: 49–61.

Savage-Rumbaugh, E. S. and Rumbaugh, D. M. and Fields, W. M. 2006. Language as a Window on Rationality. In: *Rational Animals?*, eds. S. Hurley and M. Nudds, 513–52. Oxford: University Press.

Savage-Rumbaugh, E. S. and Rumbaugh, D. M. and McDonald, K. 1985. Language Learning in Two Species of Apes. *Neuroscience and Behavioral Reviews* 9: 653–65.

Savage-Rumbaugh, E. S. and Scanlon, J. L. and Rumbaugh, D. M. 1980. Communicative Intentionality in the Chimpanzee. *Behavioral and Brain Sciences* 3: 620–23.

Savage-Rumbaugh, E. S. and Shanker, S. G. and Taylor, T. J. 1998. *Ape Language and the Human Mind*. Oxford: Oxford University Press.

Savage-Rumbaugh, E. S. *et al.* 1985. The Capacity of Animals to Acquire Language. *Philosophical Transactions of the Royal Society B* 308: 177–85.

Savage-Rumbaugh, E. S. *et al.* 1986. Spontaneous Symbol Acquisition and Communicative Use by Pigmy Chimpanzees (*Pan Paniscus*). *Journal of Experimental Psychology: General* 115: 211–35.

Savage-Rumbaugh, E. S. *et al.* 1990. Symbols. In: *Advances in Infancy Research*, Vol. 6, eds. C. Rovee-Collier and L. P. Lipsitt, 221–78. E. Norwood (NJ): Ablex.

Savage-Rumbaugh, E. S. *et al.* 1993. Language Comprehension in Ape and Child. *Monographs of the Society for Research in Child Development* 58, Serial No. 223. 1–221.

Savage-Rumbaugh, E. S. *et al.* 1996. Language Perceived. In: *Great Ape Societies*, eds. W. McGrew and L. Marchant and T. Nishida, 173–84. Cambridge, Cambridge University Press.

Savage-Rumbaugh, E. S. *et al.* 2005. Culture Prefigure Cognition in *Pan/Homo* Bonobos. *Theoria* 20: 311–28.

Schick, K. D. and Toth, N. and Garufi, G. 1999. Continuing Investigations in the Stone Tool-making and Tool-using Capabilities of a Bonobo *(Pan paniscus)*. *Journal of Archeological Science* 26: 821–32.

Schindler, S. 2007. Rehabilitating Theory. *Studies in the History and Philosophy of Science* 38: 160–84.

Schmelz, M. and Call, J. and Tomasello, M. 2011. Chimpanzees know that others make inferences. *Proceedings of the National Academy of Sciences USA* 108: 3077–3079.

Scholl, B. J. and Tremoulet, P. D. 2000. Perceptual Causality and Animacy. *Trends in Cognitive Science* 4: 299–309.

Schrödinger, E. 1926a. Quantisierung als Eigenwertproblem (Erste Mitteilung). *Annalen der Physik* 79: 361–76.

Schrödinger, E. 1926b. Quantisierung als Eigenwertproblem (Zweite Mitteilung). *Annalen der Physik* 79: 489–527.

Schrödinger, E. 1926c. Über das Verhältnis der Heisenberg-Born-Jordanschen Quantenmechanik zu der meinem. *Annalen der Physik* 79: 734–56.

Schrödinger, E. 1926d. Quantisierung als Eigenwertproblem (Dritte Mitteilung). *Annalen der Physik* 80: 85–138.

Schrödinger, E. 1926e. Quantisierung als Eigenwertproblem (Vierte Mitteilung). *Annalen der Physik* 81: 109–39.

Schrödinger, E. 1935. Die gegenwärtige Situation in der Quantenmechanik *Naturwissenschaften* 23: 807–12, 823–28, 844–49.

Schuster, A. 1883. The Genesis of Spectra. *Report of the Fifty-Second Meeting of the British Association for the Advancement of Science*: 120–49.

Schutz, A. 1954. Concept and Theory Formation in the Social Sciences. *Journal of Philosophy* 51: 257–73.

Sevcik, R. and Savage-Rumbaugh, E. S. 1994. Language Comprehension and Use by Great Apes. *Language and Communication* 14: 37–58.

Shapere, D. 1982. The Concept of Observation in Science and Philosophy. *Philosophy of Science* 49: 485–525.

Sherrow, M. M. 2005. Tool use in insect foraging by the chimpanzees of Ngogo, Kibale National Park, Uganda. *American Journal of Primatology* 65: 377–83.

Shultz, S. and Christopher, O. and Atkinson, Q. D. 2011. Stepwise Evolution of Stable Sociality in Primates. *Nature* 479, No. 7372: 219–22.

Sibum, H. O. 1995. Reworking the Mechanical Value of Heat. *Studies in the History and Philosophy of Physics* 26: 73–106.

Sintonen, Matti. 2005. Scientific Explanation. *Synthese* 143: 179–205.

Skinner, B. F. 1957. *Verbal Behavior*. New York: Appleton-Century-Crofts.

Smith, A. 1981. *An Inquiry into the Nature and Causes of the Wealth of Nations*, Vol. I. Glasgow Edition of the Works and Correspondence Vol. 2a. Indianapolis (IN): Liberty Fund.

Smith, C. W. 1977. William Thomson and the Creation of Thermodynamics. *Archive for the History of Exact Sciences* 16: 213–88.

Sommerfeld, A. 1922. *Atombau und Spektrallinien*. Braunschweig: Vieweg und Sohn.

Sommerfeld, A. 1923. *Atomic Structure and Spectral Lines*. London: Methuen and Co.

Sommerfeld, A. 1927. Zum gegenwärtigen Stande der Atomphysik. *Zeitschrift für Physik* 28: 231–39.

Stevens, J. R. 2004. The selfish nature of generosity: harassment and food sharing in primates. *Philosophical Transactions of the Royal Society B* 271: 451–56.

Stoney, J. G. 1868. On the Internal Motions of Gases Compared with the Motions of Waves of Light. *Philosophical Magazine* 36: 132–41.

Stoney, J. G. 1871. On the Cause of the Interrupted Spectra of Gases. *Report of the Fortieth Meeting of the British Association for the Advancement of Science*: 41–3.

Stoney, J. G. 1872. On the Advantage of Referring the Positions of Lines in the Spectrum to a Scale of Wave-numbers. *Report of the Forty-First Meeting of the British Association for the Advancement of Science*: 42–3.

Stoney, J. G. 1891. On the Cause of Double Lines and of Equidistant Satellites in the Spectra of Gases. *Scientific Transactions of the Royal Society of Dublin* 4: 563–608.

Stoney, J. G. 1894. Of the "Electron" or Atom of Electricity. *Philosophical Magazine* 38: 418–20.

Strevens, M. 2007. Review of Woodward, "Making Things Happen". *Philosophy and Phenomenological Review* 74: 233–49.

Strevens, M. 2008. Comments on Woodward's "Making Things Happen". *Philosophy and Phenomenological Review* 77: 171–92.

Strutt, J. W. (Baron Rayleigh) 1906. On Electrical Vibrations and the Constitution of the Atom. *Philosophical Magazine* 11: 117–23. Reprinted in: *Scientific Papers*, Vol. 5. Cambridge 1912, Cambridge University Press: 287–91.

Styer, D. F. 1996. Common Misconceptions Regarding Quantum Mechanics. *American Journal of Physics* 64: 31–4.

Such, J. 1978. Idealization and Concretization in the Natural Sciences. In: *Aspects of the Growth of Science*, ed. W. Krajewski, 49–73. Amsterdam: Grüner.

Taglialatela, J. P. and Savage-Rumbaugh, E. S. and Baker, L. A. 2003. Vocal Production by a Language-Competent Bonobo (*Pan Paniscus*). *International Journal of Primatology* 24: 1–17.

Tedlock, B. 1991. From Participant Observation to the Observation of Participation: The Emergence of Narrative Ethnography. *Journal of Anthropological Research* 47: 69–94.

Tedlock, B. 2004. Narrative Ethnography as Social Science Discourse. *Studies in Symbolic Interaction* 27: 23–31.

Tennie, C. and Call, J. and Tomasello, M. 2009. Ratcheting Up the Ratchet on the Evolution of Cumulative Culture. *Philosophical Transactions of the Royal Society B* 364: 2405–2415.

Thomson, J. J. 1897. On Cathode Rays. *Proceedings of the Cambridge Philosophical Society* 9: 243–44.

Thomson, W. (Lord Kelvin) 1848. On an Absolute Thermometric Scale Founded on Carnot's Theory of the Motive Power of Heat. *Philosophical Magazine* 33: 313–17.

Thomson, W. (Lord Kelvin). 1849. An Account of Carnot's Theory of the Motive Power of Heat. *Transactions of the Royal Society of Edinburgh* 16: 541–74.

Thomson, W. (Lord Kelvin). 1853a. On the Dynamic Theory of Heat. *Transactions of the Royal Society of Edinburgh* 20: 261–88.

Thomson, W. (Lord Kelvin). 1853b. On a Method of Discovering Experimentally the Relation between the Mechanical Work Spent, and the Heat Produced by the Compression of a Gaseous Fluid. *Transactions of the Royal Society of Edinburgh* 20: 289–316.

Thomson, W. (Lord Kelvin). 1853c. On the Dynamic Theory of Heat, Part V. *Transactions of the Royal Society of Edinburgh* 20: 475–82.

Tomasello, M. 1990. Cultural Transmission in the Tool Use and Communicatory Signalizing of Chimpanzees? In: *Language and Intelligence in Monkeys and Apes*, eds. S. T. Parker and K. C. Gibson, 274–311. Cambridge: Cambridge University Press.

Tomasello, M. 1996. Do Apes Ape? In: *Social Learning in Animals*, eds. B. Galef and C. Heyes, 319–46. New York: Academic Press.

Tomasello, M. 1999a. *Cultural Origins of Human Cognition*. Cambridge (MA): Harvard University Press.

Tomasello, M. 1999b. The Human Adaptation for Culture. *Annual Review of Anthropology* 28: 509–29.

Tomasello, M. 2000. Culture and Cognitive Development. *Current Directions in Psychological Science* 9: 37-40.

Tomasello, M. 2003. *Constructing a Language.* Cambridge (MA): Harvard University Press.

[Michael Tomasello's interview on 3SAT television station] accessible on http://video.google.com/videoplay?docid=8933367116959974563#]

Tomasello, M. 2008. *The Origins of Human Communication.* Cambridge (MA): MIT Press.

Tomasello, M. 2011. Human Culture in Evolutionary Perspective. In: *Advances in Culture and Psychology;* Vol. 1, eds. M. J. Gelfand and C. Chiu and Y. Hong, 5-48. Oxford: Oxford University Press.

Tomasello, M. 2012. The Ultra-social Animal. *European Journal of Social Psychology* 44: 187-94.

Tomasello, M. 2014. *The Natural History of Human Thinking.* Cambridge (MA): Harvard University Press.

Tomasello, M. and Call, J. 1994. Social Cognition of Monkeys and Apes. *Yearbook of Physical Anthropology* 37: 273-305.

Tomasello, M. and Call, J. 1997. *Primate Cognition.* Oxford: Oxford University Press.

Tomasello, M. and Call, J. 2006. Do Chimpanzees Know What Others See or Only What They Are Looking At? In: *Rational Animals?* eds. S. Hurley and T. Nudds, 371-84. Oxford: Oxford University Press.

Tomasello, M. and Carpenter, M. 2005a. The Emergence of Social Cognition in Three Young Chimpanzees. *Monographs of the Society for Research in Child Development* 58, Serial Number 70: 1-28.

Tomasello, M. and Carpenter, M. 2005b. The Emergence of Social: A Longitudinal Study. *Monographs of the Society for Research in Child Development* 58, Serial Number 70: 29-45.

Tomasello, M. and Carpenter, M. 2005c. General Discussion. *Monographs of the Society for Research in Child Development* 58, Serial Number 70: 107-22.

Tomasello, M. and Carpenter, M. 2007. Shared Intentionality. *Developmental Science* 10: 121-25.

Tomasello, M. and Moll, H. 2010. The Gap is Social. In: *Mind the Gap,* eds. P. M. Kappeler and Silk, 331-49. Heidelberg: Springer.

Tomasello, M. and Moll, H. 2013. Why don't Apes Understand False Beliefs? In: *Navigating the Spatial World*, eds. M. R. Banaji and S. A. Gelman, 81–88. Oxford: Oxford University Press.

Tomasello, M. and Rakoczy, H. 2003. What Makes Human Cognition Unique? *Mind and Language* 18: 121–47.

Tomasello, M. and Call, J. and Hare, B. 2003. Chimpanzees Understand Psychological States. *Trends in Cognitive Science* 7: 153–56.

Tomasello, M. and Carpenter, M. and Liszkowski, U. 2007. A New Look at Infant Pointing. *Child Development* 78: 705–22.

Tomasello, M. and Gust, D. and Frost, G. T. 1989 A Longitudinal Investigation of Gestural Communication in Young Chimpanzees. *Primates* 30: 35–50.

Tomasello, M. and Kruger, A. and Ratner, H. 1993. Cultural Learning. *Behavioral and Brain Sciences* 16: 495–523.

Tomasello, M. et al. 1985. The Development of Gestural Communication in Young Chimpanzees. *Journal of Human Evolution* 14: 175–96.

Tomasello, M. et al. 2005. Understanding and Sharing Intentions. *Behavioral and Brain Sciences* 25: 675–745.

Tomasello, M. et al. 2012. Two Key Steps in the Evolution of Human Cooperation. *Current Anthropology* 53: 673–92.

Toth, N. et al. 1993. Pan the Tool-Maker. *Journal of Archeological Science* 20: 81–91.

Toth, N. and Schick, K. 2006. A Comparative Study of the Stone Making Skills of *Pan*, *Australopithecus*, and *Homo sapiens*. In: *The Oldowan*, eds. N. Toth and K. Schick, 155–222. Bloomington (IN): The Oldowan Stone Age Institute.

Toth, N. and Schick, K. 2009. The Oldowan. *Annual Review of Anthropology* 38: 285–305.

Toulmin, S. 1969. From Logical Analysis to Conceptual History. In: *The Legacy of Logical Positivism in Philosophy of Science*, eds. P. Achinstein and S. F. Barker, 25–53. Baltimore: Johns Hopkins Press.

Tuchańska, Barbara. 1992. What is Explained in Science? *Philosophy of Science* 59: 102–19.

Turner, D. 2005. Local Underdetermination in Historical Science. *Philosophy of Science* 72: 209–30.

Turner, D. 2007. *Making Prehistory*. Cambridge: Cambridge University Press.

Van de Bijl, H. J. 1917a. Note on the Ionizing Potential of Metallic Vapors. *Physical Review* 9: 173–75.

Van de Bijl, H. J. 1917b. Theoretical Considerations Concerning Ionization and Single-Lined Spectra. *Physical Review* 10: 546–56.

Van Fraassen, B. C. 1980. *The Scientific Image*. Oxford: Oxford University Press.

Van Vleck, J. H. 1922. The Dilemma of the Helium Atom. *Physical Review* 19: 419–20.

Van Vleck, J. H. 1923. Note on the quantum theory of the Helium arc spectrum. *Physical Review* 21: 372–73.

Visalberghi, E. and Tomasello, M. 1998. Primate Causal Understanding in the Physical and Psychological Domains. *Behavioral Record* 42: 189–203.

Vogel, H. W. 1880. Über die Spectra des Wasserstoffs, Quecksilbers und Stickstoffs. *Monatsberichte der königlichen Akademie der Wissenschaften zu Berlin*, 10. Juli 1879. Berlin: Verlag der königlichen Akademie der Wissenschaften: 586–609.

Von Helmholtz, H. 1880. Über die photographische Aufnahme von Spectren der in Geisslerröhren eingeschlossenen Gase. *Monatsberichte der Königlichen Preußischen Akademie der Wissenschaften zu Berlin*, 1879: 115–19.

Walsh, W. H. 1953. Categories. *Kant Studien* 45: 274–85.

Watkins, J. W. N. 1958. Confirmable and Influential Metaphysics. *Mind* 67: 344–65.

Watkins, J. W. N. 1975. Metaphysics and the Advancement of Science. *British Journal for the Philosophy of Science* 26: 91–121.

Wayne, A. 2011. Expanding the Scope of Explanatory Idealization. *Philosophy of Science* 78: 830–41.

Weatherson, B. 2012. Explanation, Idealisation and the Goldilocks Problem. *Philosophy and Phenomenological Research* 84: 461–73.

Weisberg, M. 2007. Three Kinds of Idealization. *Journal of Philosophy* 104: 639–59.

Whewell, W. 1834. On the Nature of the Truth of the Laws of Motion. *Transactions of the Cambridge Philosophical Society* 5: 149–72.

Wilcke, J. C. 1772a. Om snöns kyla vid smältningen. *Kongliga vetenskaps academien handlingar* 33: 97–120.

Wilcke, J. C. 1772b. Von des Schnees Kälte beym schmelzen. *Der königlichen schwedischen Akademie der Wissenschaften Abhandlungen* 34: 93–116.

Wilcke, J. C. 1781a. Rön, Om eldens specifica myckenhet uti fasta kroppar, och des afmätande. *Kongliga vetenskaps academien nye handlingar* 2: 49–78.

Wilcke, J. C. 1781b. Über die spezifische Menge des Feuers im festen Körpern und derselben Abmessung. *Der königlichen schwedischen Akademie der Wissenschaften neue Abhandlungen* 2: 48–79.

Wilfried, E. E. G. and Yamagiwa, J. 2014. Use of tools sets by chimpanzees for multiple purposes in Moukalaba-Doudou National Park, Gabon. *Primates* 55: 467–72.

Woodward, J. 1979. Scientific Explanation. *British Journal for the Philosophy of Science* 30: 41–67.

Woodward, J. 1980. Developmental Explanation. *Synthese* 44: 443–66.

Woodward, J 1984. A Theory of Singular Causal Explanation. *Erkenntnis* 21: 231–62.

Woodward, J. 1989. Data and Phenomena. *Synthese* 79: 393–472.

Woodward, J. 1997. Explanation, Invariance and Intervention. *Philosophy of Science* 64 (Proceedings): S26–S41.

Woodward, J. 2000a. Data, Phenomena, and Reliability. *Philosophy of Science* 67: S163 – S179.

Woodward, J. 2000b. Explanation and Invariance in Special Science. *British Journal for the Philosophy of Science* 51: 197–254.

Woodward, J. 2003. *Making Things Happen*. Oxford: Oxford University Press.

Woodward, J. 2011. Data and Phenomena. *Synthese* 182: 165–79.

Woodward, J. and Hitchcock, C. 2003. Explanatory Generalizations, Part I. *Noûs* 37: 1–24.

Ylikoski, P. and Kuorikoski, J. 2010. Dissecting Explanatory Power. *Philosophical Studies* 148: 201–19.

www.ingramcontent.com/pod-product-compliance
Ingram Content Group UK Ltd.
Pitfield, Milton Keynes, MK11 3LW, UK
UKHW041902230426
12049UKWH00002B/18